Global Symposium on Leadership and Project Management 2023

Online
24-25 August 2023

ISBN: 978-1-7138-7680-9
eISBN: 978-1-7138-7712-7

Printed from e-media with permission by:

Curran Associates, Inc.
57 Morehouse Lane
Red Hook, NY 12571

Phone: (857) 301-0010

pmsymposium@cps.northeastern.edu

Additional copies of this publication are available from:

Curran Associates, Inc.
57 Morehouse Lane
Red Hook, NY 12571 USA
Phone: 845-758-0400
Fax: 845-758-2633
Email: curran@proceedings.com
Web: www.proceedings.com

TABLE OF CONTENTS

A Survey Study on Barriers and Incentives to Adoption of Sand Substitutes

Adel A. Zadeh[1*], Yunxin Peng[2] and Sheila M. Puffer[3]
[1]College of Professional Studies, Northeastern University, Toronto, ON M5X IE2, Canada (a.zadeh@northeastern.edu)
[2]D'Amore-McKim School of Business, Northeastern University, Boston, MA 02115, USA (peng.yunx@northeastern.edu)
[3]D'Amore-McKim School of Business, Northeastern University, Boston, MA 02115, USA (s.puffer@northeastern.edu)

Abstract

The escalating global sand crisis has garnered significant attention in recent years, prompting the United Nations to put forth initiatives aimed at limiting the reliance on river and coastal sand. However, despite the availability of several promising sustainable alternatives to alluvial sand, the construction industry has shown limited commitment to implementing these proposed measures. The primary objective of this article is to foster a more comprehensive understanding of the barriers, challenges, and incentives associated with the adoption of sustainable alternatives to sand in the construction industry. The authors developed and disseminated a carefully designed online survey to solicit the perspectives of construction industry professionals situated across 35 states in the United States and 7 provinces in Canada. The analysis of responses from 344 participants revealed divergent priorities and concerns among various construction stakeholders regarding sustainable sand substitutes. While some emphasized technical and practical aspects, others placed greater emphasis on long-term environmental considerations. Encouragingly, the study identified that incentives such as tax credits, subsidies, and research grants were viewed as having the potential to encourage greater adoption of sustainable sand materials and practices. The study also identified a dearth of industry-wide standards and insufficient awareness regarding the sand crisis as key factors contributing to the lack of urgency among construction industry stakeholders. These findings call for public policy measures to promote widespread adoption of sustainable sand practices in construction. Policymakers, industry stakeholders, and others can drive change by raising awareness, providing incentives, implementing regulations, and fostering regional solutions for sustainable sand substitutes.

Keywords: Sand crisis, construction industry, sustainable practices, incentives, barriers

1. Introduction

Over the past few years, the global sand crisis has garnered considerable attention from environmental groups. The United Nations has emphasized the need for reduced sand consumption and has proposed solutions to address this crisis, such as adopting sustainable substitutes for sand (UNEP, 2019, 2022). Despite the perception of sand as an abundant resource, it is actually finite

https://doi.org/10.52202/070182-0001

and increasingly scarce, playing a vital role in industries like construction, manufacturing, and energy production (Cao & Masanet, 2022). However, the extraction and utilization of sand have profound impacts on the environment and society, leading to habitat destruction, water pollution, and coastal erosion (UNEP, 2019, 2022). To mitigate these negative consequences and ensure the long-term viability of this essential resource, it is imperative to embrace more sustainable substitutes for sand as the demand continues to rise (Torres et al., 2017). The sand crisis has wide-ranging environmental and societal consequences (Wesley & Puffer, 2018). Extraction causes habitat degradation and air pollution during transportation (UNEP, 2022). Demand-driven conflicts over sand access give rise to illegal extraction by powerful sand mafias (Beiser, 2017).

Extensive research has been devoted to developing sustainable alternatives to traditional sand in concrete applications (Zadeh et al., 2022). However, limited knowledge exists regarding the challenges hindering their widespread application and the incentives necessary to promote their use. Thus, this research aims to identify the primary barriers to adopting sustainable options and the incentives that could facilitate their wider adoption. Understanding these challenges and barriers can guide the development of strategies and approaches to promote sustainable practices. Furthermore, this research can inform policy decisions, shape regulations, and advocate for responsible sand use, thereby contributing to responsible resource utilization. Overall, this study holds significant potential in advancing sustainable practices during the sand crisis. The following sections include a literature review on barriers to adopting sustainable building materials, details on the survey methodology, results and discussion, future research directions, and a conclusion.

2 Literature Review

To address the pressing issue of the sand crisis in a responsible and sustainable manner, it is essential to embrace more sustainable materials (Cao & Masanet, 2022; UNEP, 2022). However, despite the urgency for action, significant barriers hinder the widespread implementation of initiatives in this industry. Moreover, the incentives that could promote the use of alternative materials are not yet clear. The next section provides an overview of the primary obstacles and incentives to adopting sustainable materials and practices in the construction industry as a whole, with a specific focus on the sand crisis and the utilization of environmentally friendly building materials.

2.1 Barriers to the Adoption of Sustainable Building Materials

Barriers to the adoption of sustainable building materials in the construction industry can be grouped into three main categories: lack of information, education, and research; cost concerns; and lack of incentives or support. One of the main obstacles is the lack of information and awareness among stakeholders in the construction industry regarding the sand crisis and available alternatives (Puffer et al., 2022; Zadeh et al., 2022). Insufficient education and training further contribute to a lack of enthusiasm and commitment towards sustainable substitutes (Ikediashi et al., 2012; Karji et al., 2020; Marker et al., 2014). The presence of technical support and expertise is also crucial for successful adoption but often insufficient (Karji et al., 2020). Addressing these barriers can be achieved through education and awareness campaigns that increase public

https://doi.org/10.52202/070182-0001

understanding and empower individuals to make informed decisions (AlSanad, 2015; Zainul Abidin, 2010).

Cost concerns pose another significant challenge in sustainability adoption in the construction industry (Högberg, 2014; Marker et al., 2014; Schmidt & Osebold, 2017; Sourani & Sohail, 2011; Zulu et al., 2022). Sustainable materials are often more expensive than traditional options (Sourani & Sohail, 2011). Factors such as advanced extraction processes, lack of economies of scale, and price transparency issues contribute to the higher cost (Martek et al., 2019; Zadeh et al., 2022). The additional expenses make it difficult for practitioners and policymakers to justify the use of sustainable materials, particularly when budgets are limited (Balasubramanian & Shukla, 2017; Sourani & Sohail, 2011).

A number of studies have also reported the lack of incentives or support as one of the main factors hindering sustainability adoption (Högberg, 2014; Olanipekun et al., 2016). Insufficient financial incentives, such as subsidies or tax credits, make it challenging for stakeholders to justify the higher cost of sustainable materials (Olanipekun et al., 2016). Similarly, the absence of regulatory support and standards can impede their adoption (Dewick & Miozzo, 2002; Gan et al., 2015; Oke & Aigbavboa, 2017; Revell & Blackburn, 2007; Yin et al., 2018). Governments, organizations, and industry groups can play a vital role in providing incentives, technical support, and implementing regulations to promote the use of sustainable materials (Tokbolat et al., 2020; Williams & Dair, 2007; Zulu et al., 2022).

The lack of interest and demand from clients and customers is another barrier. Limited awareness of the benefits of sustainable materials, perceptions of their inferior quality, and concerns about their cost can result in a lack of demand (Costa et al., 2018; Zadeh et al., 2022). Manufacturers and suppliers may hesitate to invest in sustainable materials without sufficient client demand, leading to limited availability in the market (Kappenthuler & Seeger, 2020). Increasing sustainability awareness and highlighting the benefits of sustainable materials is essential to drive interest and demand (Häkkinen & Belloni, 2011).

2.2 Incentives to the Adoption of Sustainable Building Materials

This study examines various incentives for promoting sustainability adoption in the construction industry across diverse applications. Specifically, our focus centers on economic incentives and regulatory incentives, as they are deemed highly relevant to sustainability initiatives when applied to construction materials.

Economic Incentives

Economic incentives, including subsidies, tax credits, grants, and other financial rewards, are commonly employed by governments to encourage the adoption of sustainable construction materials and practices (Agyekum et al., 2021; Liu & Xu, 2015; Pitt et al., 2009). These incentives play a crucial role in increasing the affordability of green construction, thereby driving the adoption of sustainable practices and materials (Pitt et al., 2009). Several studies have demonstrated the effectiveness of economic incentives in promoting sustainable initiatives within the construction industry (Olanipekun et al., 2016; Rana et al., 2021). By offering tax credits, grants, and other financial rewards, both government and private sectors can provide the necessary

https://doi.org/10.52202/070182-0001

incentives for industry-wide adoption (Liu & Xu, 2015). Furthermore, economic analysis indicates that subsidies and other economic incentives can help reduce the costs associated with sustainable building materials, thus encouraging their widespread adoption in construction projects (Li et al., 2018).

Liu & Xu (2015) conducted a study analyzing the green building incentive policies in western countries and China. They emphasized the importance of implementing more efficient and effective incentives to drive the growth of the green building market. Li et al. (2018) examined the impact of government-affiliated green loans and subsidies on clean production. Their findings demonstrated that government subsidies can effectively reduce energy emissions and improve environmental quality. However, they also noted that subsidies can increase innovation risks while raising enterprise enthusiasm for innovation. Berawi et al. (2020) explored the role of green building concepts in promoting sustainable construction in developing countries. They suggested that encouraging incentive models, such as streamlined licensing and certification processes and reduced property taxes, can effectively support sustainable construction initiatives.

In another study Saka et al. (2021) conducted an in-depth analysis of voluntary incentives provided by governments in the construction industry. They proposed that scaling these incentives according to project phases can enhance stakeholder participation. Okoye (2021) assessed the influence of factors of sustainable construction on public and private sector clients and found that client knowledge, cost implications, and economic value were key drivers of commitment to sustainability. The effectiveness of economic incentives in promoting sustainability in the construction industry depends on the design and goals of the incentive package (Saka et al., 2021). The incentives should be attractive and accessible, offering financing options and rewards for achieving milestones (Rana et al., 2021). Proper communication and marketing of the incentives to the construction industry are also crucial for companies to be aware of and take advantage of them (Agyekum et al., 2021; Bond, 2011). In summary, economic incentives can be effective in promoting sustainability in the construction industry, provided they are appropriately designed and implemented.

Regulatory Incentives

Regulatory incentives encompass laws, regulations, and standards that encourage or mandate the adoption of sustainable construction practices (Dewick & Miozzo, 2002; Winston, 2010). These incentives are established in response to global environmental crises, such as the sand crisis, with the aim of mitigating the negative impacts of unsustainable practices (UNEP, 2022). They have been recognized as a powerful driver of sustainable construction practices (Yudelson, 2008). As green building practices gain traction worldwide, the number of regulations incentivizing or mandating them is increasing (Andelin et al., 2015). Governments are actively promoting green building practices and energy efficiency through legislative measures, executive orders, and national policies, such as the Customer Choice and Electricity Reliability Act and the Clean, Renewable, and Efficient Energy Act in Michigan (Mulligan et al., 2014). Similarly, states like Washington and California have stringent local codes and regulations to incentivize stakeholders to adopt green practices despite perceived higher costs (Korkmaz et al., 2010). Countries in Asia, including China, Singapore, Hong Kong, and India, have also embraced the concept and

https://doi.org/10.52202/070182-0001

implemented regulatory incentives (Gou & Lau, 2014). As the green building movement continues to expand, it is anticipated that the number of regulatory requirements will increase accordingly. However, the construction industry often faces resistance to regulatory incentives due to the associated compliance costs (Okoye, 2021; Saka et al., 2021). It is crucial to ensure effective enforcement of regulations and hold non-compliant construction firms accountable (Chan et al., 2018; Yuanyuan Li et al., 2022). Nevertheless, properly designed and implemented regulatory incentives can be effective. Awareness of existing regulations and the potential consequences of non-compliance is essential for successful implementation of sustainability initiatives in the construction industry. In summary, similar to economic incentives, regulatory incentives hold potential for driving sustainable construction practices, requiring industry awareness and compliance to realize their benefits.

3 Methods

3.1 Overview

In this study, we employed the stakeholder opinion poll pyramid (SOPP) method, which has been utilized in previous studies (Hugé et al., 2010; Kamari et al., 2017; Tokbolat et al., 2020). This method enables a systematic approach to gather and prioritize stakeholder opinions, facilitating a comprehensive evaluation of important factors. The procedural steps involved the following sequence: (1) Extraction of sustainability drivers and barriers from the existing literature; (2) Preselection, sorting, and analysis to identify the parameters to be evaluated by experts; (3) Collection of stakeholder evaluations on each parameter using a structured survey with a quantitative scale; (4) Consolidation of all gathered information.

Initially, potential sustainability barriers and incentives in the construction industry were identified through an extensive literature review. These identified items were then validated by industry experts through one-hour interviews, assessing their relevance and importance. To investigate the perspectives of key stakeholders in the construction industry regarding incentives for adopting sustainable sand substitutes, a survey was designed as the primary method of data collection. Further explanations regarding the survey design, distribution, and data analysis approach will be provided in the subsequent sections.

3.2 Survey Instrument and Data Collection

This study was part of a larger research program exploring the adoption of sustainable sand substitutes in the construction industry. A survey was designed using Qualtrics and piloted with professionals in the construction field. From May 2020 to June 2021, the survey was administered to professionals across the United States and Canada through various recruitment methods. In total, 337 usable responses were collected. The focus on these two countries was due to their significant construction sectors and potential for enacting legislation to promote sustainable substitutes. Participants had the opportunity to receive four $50 gift cards as incentives for their participation.

The survey participants were asked to choose three attributes that they value in sustainable sand replacement materials and rank them according to importance, using a scale of 1 to 3. The attributes provided were listed in alphabetical order and included availability or ease of purchase, customer demand, environmental friendliness, performance (such as durability, strength, and permeability), price, regulatory approvals of use, and not enough knowledge to answer.

https://doi.org/10.52202/070182-0001

Additionally, respondents had the option to add any attributes that were not included in the provided options. Similarly, participants were presented with a list of concerns about sustainable sand replacement materials, also in alphabetical order. The options included availability and ease of purchase, customer demand, capital investment required for sand substitute infrastructure, environmental friendliness, performance (including durability, strength, and permeability), price, regulatory approvals of use, sunk costs in existing sand manufacturing infrastructure, not enough knowledge to answer, and the opportunity to include any concerns that were not covered in the given options

The survey included three questions to assess respondents' views on incentives and their effectiveness in promoting the adoption of sustainable substitutes for sand. In the first question, respondents were asked to select up to three incentives relevant to their organization that would encourage them to switch to sand substitutes. They were then asked to rank these incentives in order of importance, with 1 being the most important and 3 being the third most important. The incentives provided were listed alphabetically and included Availability/Convenience to Purchase, Financial Benefit, Information Available, Protection of the Environment, Requests from Clients, Specific Regulations to be Followed, Social Responsibility, Technical Support, Training Programs, and Not Enough Knowledge to Answer. Respondents also had the opportunity to suggest additional incentives. In the second question, respondents were presented with various governmental incentives related to sustainable sand replacement materials. They were asked to rank their top three choices from the options provided, listed alphabetically: convince companies to take climate issues seriously, develop voluntary guidelines (e.g., LEED), financial incentives, impose mandatory regulations, invest in R&D, issue government sustainability action plans, non-monetary incentives such as repeat business, sponsor public information campaigns, not enough knowledge to answer, and the option to add any incentives that were not included. The third question focused specifically on financial incentives. Respondents were asked to rank their top three choices from the following options: funding grant program, loan and mortgage program, property tax abatement, rebates for adoption, repeat business, tax credit, tax increment financing reimbursement, not enough knowledge to answer, and the opportunity to suggest additional financial incentives.

4 Results and Discussion

The demographic data collected from survey participants aimed to provide insights into the sample composition and compare responses among five stakeholder groups. The majority of respondents came from medium-sized organizations with 50-250 employees, had 6-15 years of experience, and fell within the age range of 30-49. The majority of respondents were male (67.06%) and most were based in the United States (69.48%). Engineers (31.11%) and managers (29.60%) were the most represented roles, followed by academics (21.14%), architects (12.99%), and a mix of other professionals (5.13%) including sustainability consultants, trade association representatives, government officials, and NGO members. A summary of the key demographic data can be found in Table 1 and Figure 1.

Table 1 Demographic data of respondents

Demographic data	Value	Number of participants (n)

Country	Canada	101
	US	230
Gender	Female	102
	Male	222
	Other/Non-Response	7
Roles	Architect	43
	Engineer	103
	Academics	70
	Management	98
	Others	17

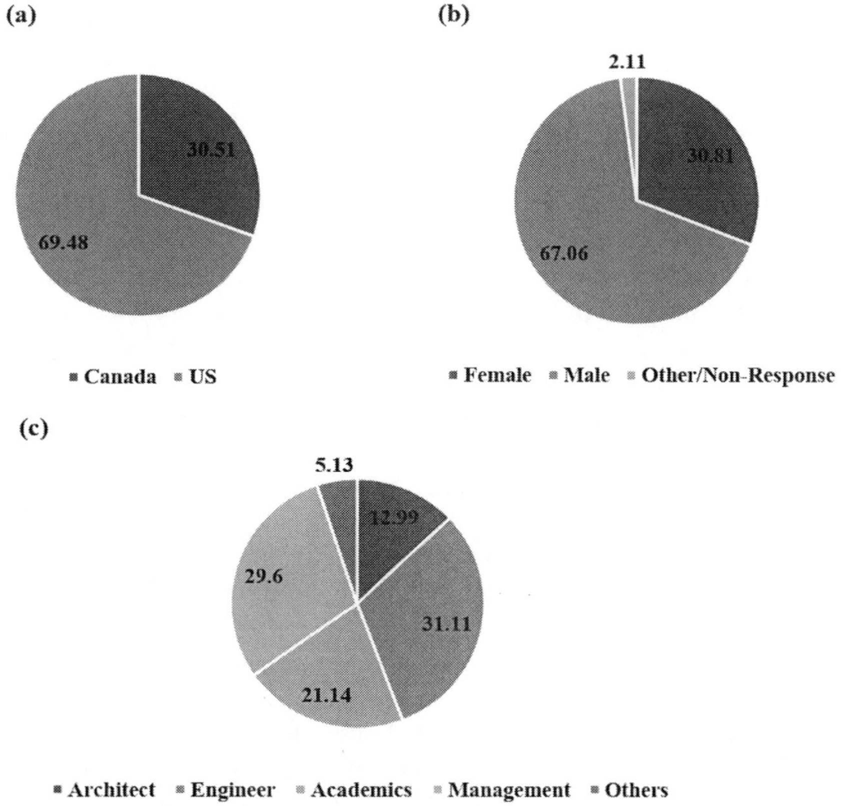

Figure 1 Demographic data of respondents (%)

4.1 Barriers and Attributes

Figure 2 displays the top three attributes and concerns selected by different professional roles in the construction industry regarding sand sustainability. With regards to attributes, Figure 2(a), the findings reveal that environmental friendliness, performance, and availability/ease of use are highly regarded by key stakeholders in the industry. These attributes were consistently chosen as the top three priorities by various groups, such as architects, academics, and engineers, indicating their widespread recognition and significance within the field. Architects' prioritization of

https://doi.org/10.52202/070182-0001

"environmental friendliness" as one of their top attributes is understandable given their responsibility for sustainable and energy-efficient building design and construction (Puffer et al., 2022; Zadeh et al., 2022). They are likely to be mindful of the environmental impact of materials and, therefore, prioritize environmentally friendly options (De Gaulmyn & Dupre, 2019). Engineers and academics also prioritized "environmental friendliness," which can be attributed to their understanding of the environmental impact of materials and their preference for those with lower environmental footprints (Kevern, 2011; Lewis, 2004; Weisenberger, 2011). Additionally, performance and price are important considerations for these groups due to their responsibility for ensuring material suitability and cost-effectiveness (Nasereddin & Price, 2021). Figure 2(a) also indicates that management prioritized performance, price, and availability/ease of use as their top three attributes. This aligns with their role in overseeing the construction process and ensuring materials are readily accessible, user-friendly, cost-effective, and perform optimally (Hinze et al., 2013).

As depicted in Figure 2(b), each stakeholder role has identified their top three concerns that shape their priorities. Architects and engineers, given their roles in construction, share similar concerns regarding sustainable sand substitutes. Both groups are responsible for designing and constructing buildings and infrastructure, making the choice of materials crucial (Puffer et al., 2022). Performance, availability/ease of use, and price are key factors influencing their material selection decisions. Performance refers to a material's ability to meet technical requirements such as strength, durability, and fire resistance, ensuring the safety and durability of structures. Availability/ease of use pertains to the convenience and accessibility of a material, which can save time and money in construction projects. Price, on the other hand, is a critical consideration for architects and engineers, as it affects the allocation of budget resources. By prioritizing performance, availability/ease of use, and price, architects and engineers can make informed decisions when it comes to sustainable sand substitutes, optimizing the design and construction process.

Academics take a broad and forward-looking perspective when it comes to sand sustainability. Their first concern is tied between capital investment and environmental friendliness, emphasizing their focus on sustainable development and the future. They recognize that investing in sustainable materials now can have long-term benefits for the environment and society. Performance is their second concern, indicating their understanding of the technical aspects of materials. Additionally, academics consider price, customer demand, and availability/ease of use as their third-level concerns, reflecting their recognition of market viability, practicality, and accessibility. They acknowledge that materials need to be economically feasible, marketable, and easy to obtain and work with for widespread adoption. The concerns expressed by academics regarding sand sustainability highlight their holistic approach, considering the technical, practical, and long-term environmental aspects. They understand that achieving sustainable development requires balancing economic, technical, and environmental considerations, emphasizing the need for a comprehensive approach to address the challenges of sand sustainability.

Management prioritizes performance as their primary concern, recognizing its crucial role in project success. Performance directly impacts safety, durability, and functionality of buildings, making it a key consideration for management. Their second concern is tied between price and availability/ease of use, underscoring their focus on cost-effectiveness and project feasibility. They understand the importance of using readily available and user-friendly materials to save time and costs. Capital investment is their third-level concern, indicating their awareness of the long-term

https://doi.org/10.52202/070182-0001

financial implications and the potential environmental and societal benefits of investing in sustainable materials. In summary, management's concerns regarding sustainable sand substitutes highlight their emphasis on the technical and practical aspects, as well as the short-term financial considerations. This demonstrates their understanding that achieving sustainable development requires a balanced approach that encompasses economic, technical, and environmental factors, emphasizing the need for a comprehensive strategy in addressing the challenges of sand sustainability.

Respondents in roles that did not align with other categories expressed their top concerns as performance, capital investment, and price. This underscores their emphasis on the technical aspects, long-term financial implications, and cost-effectiveness of the material.

Overall, the survey findings demonstrate that key stakeholders in the construction industry have diverse concerns regarding sand sustainability. Some stakeholders prioritize the technical and practical aspects, while others prioritize long-term and environmental considerations. These varying perspectives highlight the multifaceted nature of sand sustainability and underscore the importance of considering the different priorities and perspectives of stakeholders when addressing sustainable practices in the construction industry.

https://doi.org/10.52202/070182-0001

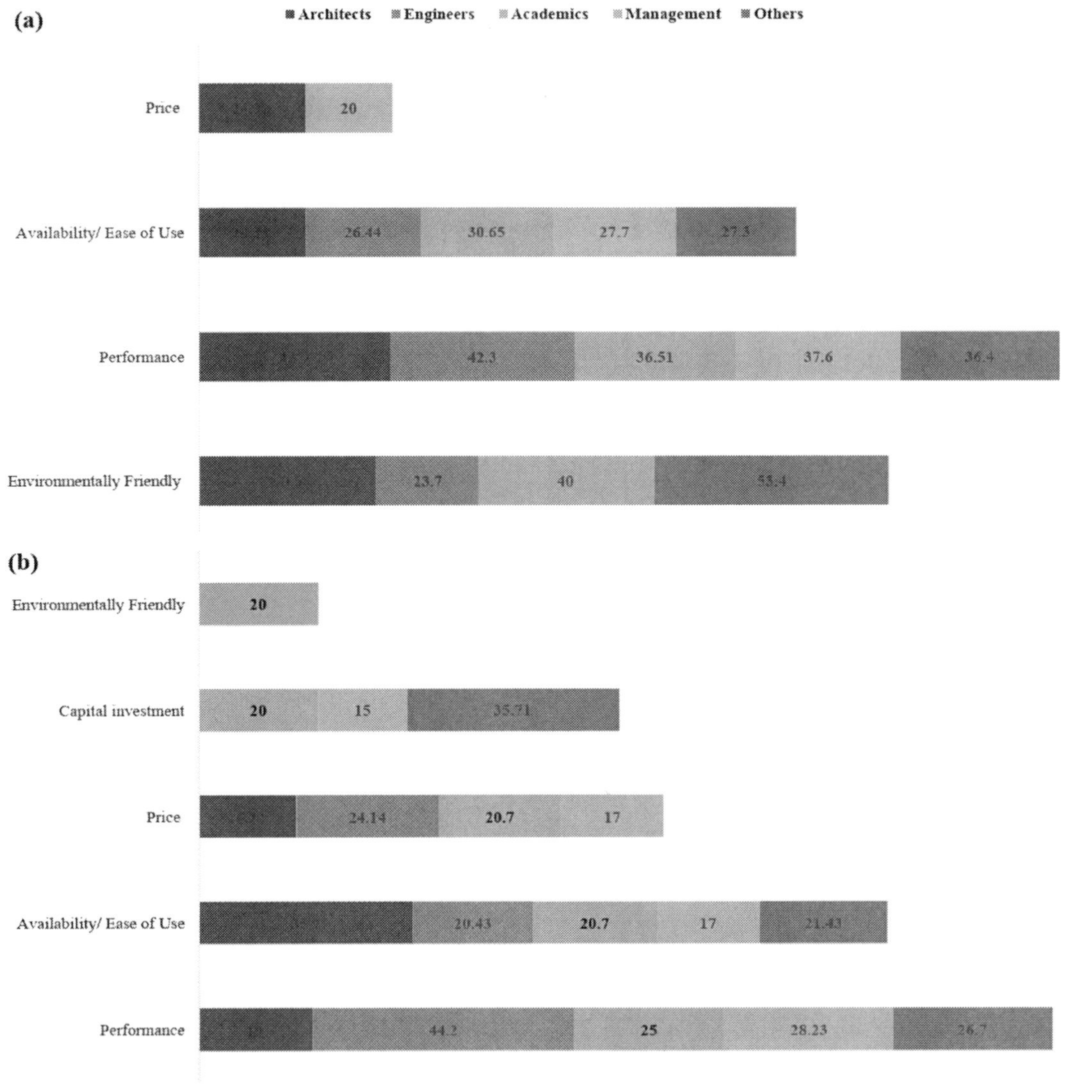

Figure 2 Top three attributes and concerns ranked by professional roles (%) (a) attributes (b) concerns

4.2 Incentives

Figure 3 displays the top three incentives (general, governmental and financial) chosen by various stakeholder roles. The percentages have been calculated based on the total number of respondents in each role and for each rank, rather than the total number of responses provided for the question.

With regards to general incentives, as shown in Figure 3(a), Protection of the environment, financial benefit, and availability/convenience of purchase were among the top three choices for each role. However, variations emerged as different professionals prioritized these incentives to varying degrees. Notably, approximately 19.05% of architects and 25% of academics ranked "Protection of the Environment" as their first option, showcasing a greater focus on sustainability

https://doi.org/10.52202/070182-0001

compared to other roles. This can be attributed to architects and academics having a deeper understanding of the environmental impacts associated with sand extraction, which is influenced by their extensive reading and research in their respective fields. As a result, they prioritize environmental protection above other factors when considering sustainable sand substitutes. Engineers and managers placed greater importance on financial benefit and availability/convenience of purchase as the primary factors influencing the adoption of sustainable sand substitutes. This emphasis can be attributed to their concerns regarding timely and cost-effective processes, which are major performance targets for them. Architects, on the other hand, highlighted technical support as their second most significant motivation for adopting sustainable sand substitutes, indicating their need for vendor assistance in making informed decisions. Engineers ranked client requests among their top three considerations, suggesting a greater inclination to follow client preferences when it comes to construction materials. Additionally, other professionals prioritized social responsibility as one of their top three incentives for adopting sustainable sand substitutes. This reflects their commitment to sustainability and their motivation to engage in environmentally friendly practices.

Figure 3(b) presents the top three governmental incentives selected by survey respondents from each professional role. The results highlight the diverse motivations among key stakeholders in the construction industry regarding the desirability of government incentives for sustainable sand substitutes. Architects prioritized the imposition of mandatory regulations (40.5%) and availability of financial incentives (32.5%) as their top choices. This suggests their interest in regulatory enforcement and financial support for adopting sustainable sand substitutes. Engineers selected financial incentives (24.2%) and mandatory regulations (21.7%), indicating their motivation to switch to sustainable alternatives with government incentives. Academics also valued financial incentives and mandatory regulations, indicating their support for government-led enforcement. Managers focused on financial incentives (28.5%) and investing in research and development (18.6%), reflecting their financial management and long-term investment concerns. Other professionals prioritized investing in research and development and developing voluntary guidelines, emphasizing their interest in the future of sustainable sand substitutes.
Overall, financial incentives were the top choice for architects, engineers, and academics. This highlights the significance of financial investment in incentivizing professionals to transition to sustainable sand substitutes. Additionally, the selection of mandatory regulations as the second choice by architects, engineers, and others indicates that a combination of financial incentives and regulations may be crucial for successful adoption.

Figure 3(c) display the top financial incentives preferred by respondents in promoting the adoption of sustainable sand substitutes. Funding grant programs and rebates for adoption were the two most desirable incentives across all stakeholder roles. Architects prioritized rebates for adoption (19.05%) as their top choice and funding grant programs (30%) as their second. Engineers and academics shared the same ranking, with funding grant programs (18.9% for engineers, 28.13% for academics) and rebates for adoption (29.1%) for engineers, 25% for academics) as their top two choices. Managers focused on funding grant programs (25.8%) and tax credits (29.9%), while other professionals prioritized funding grant programs (20%) and tax credits (18.8%). It is worth noting that a significant number of respondents selected "not enough knowledge to answer," indicating the need for further education and awareness about potential financial incentives.

https://doi.org/10.52202/070182-0001

Overall, the findings suggest that stakeholders perceive rebates for adoption, funding grant programs, and tax credits as the most effective financial incentives for promoting the adoption of sustainable sand substitutes. Therefore, governments should prioritize these types of incentives rather than loan and mortgage programs, property tax abatement, repeat business, or tax increment financing reimbursement. Such government-supported investments and incentives can reduce the cost of new technologies, making them more competitive and encouraging their adoption in the construction industry. Additionally, these incentives can provide the necessary motivation and impetus for professionals to transition to more sustainable practices and resources.

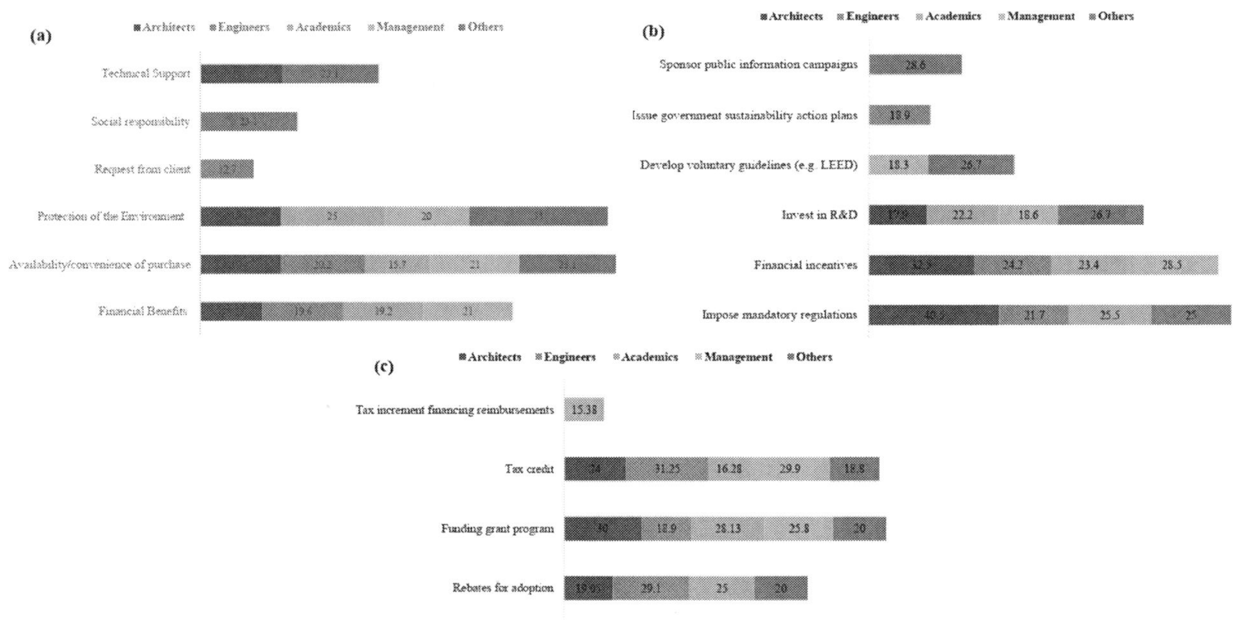

Figure 3 Top three government incentives ranked by professional roles (%) (a) general incentives (b) government incentives, (c) financial Incentives

This observation highlights the connection between the identified attributes and concerns related to sustainable sand substitutes and the corresponding expectations for solutions or incentives. It suggests that stakeholders in the construction industry are seeking specific measures or incentives to support the adoption of sustainable alternatives.

Stakeholders in the construction industry prioritize accessible, cost-effective, and environmentally friendly options for sourcing sand substitutes, reflecting their increasing awareness of environmental implications. However, the limited availability and transportation costs of these materials present challenges. To address the global sand crisis, it is crucial to maximize the use of locally-sourced sustainable alternatives and provide financial support to facilitate the adoption of such substitutes.

The survey results confirm the preference for financial incentives in promoting the adoption of sustainable sand substitutes. Stakeholders consistently selected funding grant programs, rebates for adoption, and tax credits as their top choices. This indicates a shared interest in government support and investment through these incentives. Funding grants and adoption rebates were particularly favored, showcasing stakeholders' enthusiasm for government-backed initiatives. Tax credits were also highly valued for their potential to reduce costs and enhance the competitiveness of sustainable sand substitutes. The findings of this study suggest that, as noted for other initiatives in the construction industry, a combination of financial incentives and regulations is necessary to ensure the successful adoption of sustainable sand substitutes (Berawi et al., 2020, 2020; Bubshait, 2003; Li et al., 2018; Saka et al., 2021). It is evident that stakeholders in the construction industry need to collaborate and develop incentives that effectively address their diverse needs.

The results of this study align with previous research, which found that key stakeholders prioritize financial investments and mandatory regulations (Yuanyuan Li et al., 2022). Therefore, it is crucial to implement a comprehensive framework of incentives that combines financial support and regulatory measures to encourage the adoption of sustainable sand substitutes (Darko et al., 2017). This highlights the need for further research on the role of financial incentives and regulations in the successful adoption of sustainable sand substitutes. In addition to the use of incentives, research has emphasized the importance of adopting a regional approach to effectively address the global sand crisis (UNEP, 2019, 2022). A regional perspective allows for the development and implementation of strategies that align with the specific needs and resources of a particular region. This approach enables the establishment of regulations, standards, and institutional capacity for sustainable practices within the construction industry (Berawi et al., 2020). Moreover, education and public awareness campaigns play a vital role in informing and motivating stakeholders to embrace sustainable sand substitutes (Puffer et al., 2022). By combining incentives, regional strategies, and educational initiatives, it appears possible to achieve more sustainable practices and mitigate the environmental impact of sand extraction.

5 Conclusion

This study underscores the importance of employing financial incentives and regulations to facilitate the widespread adoption of sustainable sand substitutes in the construction industry. The utilization of these incentives and regulations would encourage the incorporation of sustainable materials into construction projects and mitigate the detrimental environmental effects associated with traditional sand usage. Furthermore, the effectiveness of these measures can be enhanced through the implementation of regional strategies, educational and public awareness campaigns, and sustainability certification programs. Public-private partnerships and educational initiatives can play a pivotal role in increasing public awareness about the advantages of transitioning to sustainable alternatives, while certification programs can ensure the authenticity and quality of sustainable sand substitutes. This research makes significant contributions both theoretically and practically by shedding light on stakeholder preferences and the critical role of financial incentives and regulations in driving the adoption of sustainable sand substitutes. Through the ranking of incentives and the analysis of stakeholder perspectives, this study provides valuable insights into the key factors that influence the adoption of sustainable construction practices. From a practical standpoint, the findings of this study offer practical guidance for policymakers, industry professionals, and other stakeholders involved in the construction sector.

In conclusion, this study underscores the urgent need to address the global sand crisis and emphasizes the significance of comprehensive and effective measures to promote the widespread use of sustainable sand substitutes in the construction industry. By implementing a holistic

https://doi.org/10.52202/070182-0001

approach that combines financial incentives, regulations, regional strategies, educational campaigns, and sustainability certification programs, the industry can reduce its dependence on alluvial sand and mitigate the environmental impacts associated with its extraction and utilization.

References

Agyekum, K., Goodier, C., & Oppon, J. A. (2021). Key drivers for green building project financing in Ghana. *Engineering, Construction and Architectural Management*, *29*(8), 3023–3050. https://doi.org/10.1108/ECAM-02-2021-0131

AlSanad, S. (2015). Awareness, Drivers, Actions, and Barriers of Sustainable Construction in Kuwait. *Procedia Engineering*, *118*, 969–983. https://doi.org/10.1016/j.proeng.2015.08.538

Andelin, M., Sarasoja, A.-L., Ventovuori, T., & Junnila, S. (2015). Breaking the circle of blame for sustainable buildings–evidence from Nordic countries. *Journal of Corporate Real Estate*, *17*(1), 26–45.

Balasubramanian, S., & Shukla, V. (2017). Green supply chain management: An empirical investigation on the construction sector. *Supply Chain Management: An International Journal*, *22*(1), 58–81. https://doi.org/10.1108/SCM-07-2016-0227

Beiser, V. (2017). He who controls the sand: The mining 'mafias' killing each other to build cities. *The Guardian*, *28*.

Berawi, M. A., Basten, V., Latief, Y., & Crévits, I. (2020). Role of green building developer and owner in sustainability construction: Investigating the relationships between green building key success factors and incentives. *IOP Conference Series: Earth and Environmental Science*, *426*(1), 012061.

Bond, S. (2011). Barriers and drivers to green buildings in Australia and New Zealand. *Journal of Property Investment & Finance*, *29*(4/5), 494–509. https://doi.org/10.1108/14635781111150367

Bubshait, A. A. (2003). Incentive/disincentive contracts and its effects on industrial projects. *International Journal of Project Management*, *21*(1), 63–70. https://doi.org/10.1016/S0263-7863(01)00078-3

Cao, Z., & Masanet, E. (2022). Material efficiency to tackle the sand crisis. *Nature Sustainability*, *5*(5), Article 5. https://doi.org/10.1038/s41893-022-00869-w

Chan, A. P. C., Darko, A., Olanipekun, A. O., & Ameyaw, E. E. (2018). Critical barriers to green building technologies adoption in developing countries: The case of Ghana. *Journal of Cleaner Production*, *172*, 1067–1079. https://doi.org/10.1016/j.jclepro.2017.10.235

Costa, C., Cerqueira, Â., Rocha, F., & Velosa, A. (2018). The sustainability of adobe construction: Past to future. *International Journal of Architectural Heritage*.

Darko, A., Zhang, C., & Chan, A. P. C. (2017). Drivers for green building: A review of empirical studies. *Habitat International*, *60*, 34–49. https://doi.org/10.1016/j.habitatint.2016.12.007

De Gaulmyn, C., & Dupre, K. (2019). Teaching sustainable design in architecture education: Critical review of Easy Approach for Sustainable and Environmental Design (EASED). *Frontiers of Architectural Research*, *8*(2), 238–260.

Dewick, P., & Miozzo, M. (2002). Sustainable technologies and the innovation–regulation paradox. *Futures*, *34*(9–10), 823–840.

Gan, X., Zuo, J., Ye, K., Skitmore, M., & Xiong, B. (2015). Why sustainable construction? Why not? An owner's perspective. *Habitat International*, *47*, 61–68.

Gou, Z., & Lau, S. S.-Y. (2014). Contextualizing green building rating systems: Case study of Hong Kong. *Habitat International*, *44*, 282–289. https://doi.org/10.1016/j.habitatint.2014.07.008

Häkkinen, T., & Belloni, K. (2011). Barriers and drivers for sustainable building. *Building Research & Information*, *39*(3), 239–255.

Hinze, J., Godfrey, R., & Sullivan, J. (2013). Integration of construction worker safety and health in assessment of sustainable construction. *Journal of Construction Engineering and Management*, *139*(6), 594.

Högberg, L. (2014). *Building Sustainability: Studies on incentives in construction and management of real estate* [PhD Thesis]. KTH Royal Institute of Technology.

Hugé, J., Le Trinh, H., Hai, P. H., Kuilman, J., & Hens, L. (2010). Sustainability indicators for clean development mechanism projects in Vietnam. *Environment, Development and Sustainability*, *12*, 561–571.

Ikediashi, D. I., Ogunlana, S. O., Oladokun, M. G., & Adewuyi, T. (2012). Assessing the level of commitment and barriers to sustainable facilities management practice: A case of Nigeria. *International Journal of Sustainable Built Environment*, *1*(2), 167–176. https://doi.org/10.1016/j.ijsbe.2013.06.002

Kamari, A., Corrao, R., & Kirkegaard, P. H. (2017). Sustainability focused decision-making in building renovation. *International Journal of Sustainable Built Environment*, *6*(2), 330–350.

Kappenthuler, S., & Seeger, S. (2020). From resources to research—A framework for identification and prioritization of materials research for sustainable construction. *Materials Today Sustainability*, *7*, 100009.

Karji, A., Namian, M., & Tafazzoli, M. (2020). Identifying the Key Barriers to Promote Sustainable Construction in the United States: A Principal Component Analysis. *Sustainability*, *12*(12), Article 12. https://doi.org/10.3390/su12125088

Kevern, J. T. (2011). Green building and sustainable infrastructure: Sustainability education for civil engineers. *Journal of Professional Issues in Engineering Education and Practice*, *137*(2), 107–112.

Korkmaz, S., Riley, D., & Horman, M. (2010). Piloting evaluation metrics for sustainable high-performance building project delivery. *Journal of Construction Engineering and Management*, *136*(8), 877–885.

Lewis, M. (2004). Integrated design for sustainable buildings. *ASHRAE Journal*, *46*(9), S22.

Li, Z., Liao, G., Wang, Z., & Huang, Z. (2018). Green loan and subsidy for promoting clean production innovation. *Journal of Cleaner Production*, *187*, 421–431. https://doi.org/10.1016/j.jclepro.2018.03.066

Liu, R., & Xu, Y. (2015). Comparison of international incentive policy of green building. In E. Qi, Q. Su, J. Shen, F. Wu, & R. Dou (Eds.), *Proceedings of the 5th International Asia Conference on Industrial Engineering and Management Innovation (IEMI2014)* (pp. 319–323). Atlantis Press. https://doi.org/10.2991/978-94-6239-100-0_58

Marker, A. W., Mason, S. G., & Morrow, P. (2014). Change Factors Influencing the Diffusion and Adoption of Green Building Practices. *Performance Improvement Quarterly*, *26*(4), 5–24. https://doi.org/10.1002/piq.21160

 https://doi.org/10.52202/070182-0001

Martek, I., Hosseini, M. R., Shrestha, A., Edwards, D. J., & Durdyev, S. (2019). Barriers inhibiting the transition to sustainability within the Australian construction industry: An investigation of technical and social interactions. *Journal of Cleaner Production*, *211*, 281–292.

Nasereddin, M., & Price, A. (2021). Addressing the capital cost barrier to sustainable construction. *Developments in the Built Environment*, *7*, 100049.

Oke, A. E., & Aigbavboa, C. O. (2017). Sustainability in construction. In *Sustainable Value Management for Construction Projects* (pp. 87–106). Springer.

Okoye, P. (2021). Factors influencing clients' commitment to sustainable construction practices. *Planning*, *16*(1), 39–48.

Olanipekun, A. O., Skitmore, & Bo Xia, P. (2016). Green building incentives: A review. *Renewable and Sustainable Energy Reviews*, *59*, 1611–1621. https://doi.org/10.1016/j.rser.2016.01.028

Pitt, M., Tucker, M., Riley, M., & Longden, J. (2009). Towards sustainable construction: Promotion and best practices. *Construction Innovation*, *9*(2), 201–224. https://doi.org/10.1108/14714170910950830

Puffer, S. M., Zadeh, A. A., & Peng, Y. (2022). Awareness of the global sand crisis and sand substitutes in the construction industry in the United States and Canada: A stakeholder analysis. *International Journal of Construction Management*, *0*(0), 1–13. https://doi.org/10.1080/15623599.2022.2146277

Rana, A., Sadiq, R., Alam, M. S., Karunathilake, H., & Hewage, K. (2021). Evaluation of financial incentives for green buildings in Canadian landscape. *Renewable and Sustainable Energy Reviews*, *135*, 110199. https://doi.org/10.1016/j.rser.2020.110199

Revell, A., & Blackburn, R. (2007). The business case for sustainability? An examination of small firms in the UK's construction and restaurant sectors. *Business Strategy and the Environment*, *16*(6), 404–420.

Saka, N., Olanipekun, A. O., & Omotayo, T. (2021). Reward and compensation incentives for enhancing green building construction. *Environmental and Sustainability Indicators*, *11*, 100138.

Schmidt, J.-S., & Osebold, R. (2017). Environmental management systems as a driver for sustainability: State of implementation, benefits and barriers in German construction companies. *Journal of Civil Engineering and Management*, *23*(1), 150–162.

Sourani, A., & Sohail, M. (2011). Barriers to addressing sustainable construction in public procurement strategies. *Proceedings of the Institution of Civil Engineers-Engineering Sustainability*, *164*(4), 229–237.

Tokbolat, S., Karaca, F., Durdyev, S., & Calay, R. K. (2020). Construction professionals' perspectives on drivers and barriers of sustainable construction. *Environment, Development and Sustainability*, *22*(5), 4361–4378. https://doi.org/10.1007/s10668-019-00388-3

Torres, A., Brandt, J., Lear, K., & Liu, J. (2017). *A looming tragedy of the sand commons*. American Association for the Advancement of Science. https://doi.org/10.1126/science.aao0503

UNEP. (2019). *Sand and sustainability: Finding new solutions for environmental governance of global sand resources*. United Nations Environment Programme Geneva, Switzerland.

UNEP. (2022). *Sand and sustainability: 10 strategic recommendations to avert a crisis.* GRID-Geneva, United Nations Environment Programme, Geneva, Switzerland.

https://doi.org/10.52202/070182-0001

Weisenberger, G. (2011). Sustainability and the structural engineer. *Practice Periodical on Structural Design and Construction*, *16*(4), 146–150.

Wesley, D. T. A., & Puffer, S. M. (2018). The end of sand: Confronting one of the greatest environmental challenges of the new millennium. In *Reusable and Sustainable Building Materials in Modern Architecture* (pp. 1–27).

Williams, K., & Dair, C. (2007). What is stopping sustainable building in England? Barriers experienced by stakeholders in delivering sustainable developments. *Sustainable Development*, *15*(3), 135–147. https://doi.org/10.1002/sd.308

Winston, N. (2010). Regeneration for sustainable communities? Barriers to implementing sustainable housing in urban areas. *Sustainable Development*, *18*(6), 319–330. https://doi.org/10.1002/sd.399

Yin, B. C. L., Laing, R., Leon, M., & Mabon, L. (2018). An evaluation of sustainable construction perceptions and practices in Singapore. *Sustainable Cities and Society*, *39*, 613–620.

Yuanyuan Li, Min Li, Peidong Sang, Po-Han Chen, & Congcong Li. (2022). Stakeholder studies of green buildings: A literature review. *Journal of Building Engineering*, 104667–104667. https://doi.org/10.1016/j.jobe.2022.104667

Yudelson, J. (2008). 7—Green Building Technologies. In *Marketing Green Building Services* (pp. 126–148). Architectural Press. https://doi.org/10.1016/B978-0-7506-8474-3.50014-4

Zadeh, A. A., Peng, Y., Puffer, S. M., & Garvey, M. D. (2022). Sustainable Sand Substitutes in the Construction Industry in the United States and Canada: Assessing Stakeholder Awareness. *Sustainability*, *14*(13), Article 13. https://doi.org/10.3390/su14137674

Zainul Abidin, N. (2010). Investigating the awareness and application of sustainable construction concept by Malaysian developers. *Habitat International*, *34*(4), 421–426. https://doi.org/10.1016/j.habitatint.2009.11.011

Zulu, S. L., Zulu, E., Chabala, M., & Chunda, N. (2022). Drivers and barriers to sustainability practices in the Zambian Construction Industry. *International Journal of Construction Management*, *0*(0), 1–10. https://doi.org/10.1080/15623599.2022.2045425

 https://doi.org/10.52202/070182-0001

Enhancing Workforce Diversity: Integrating Inclusivity Parameters into Applicant Tracking Systems

Arjun Malarmannan
College of Professional Studies
Northeastern University, Boston
malarmannan.a@northeastern.edu

Abstract

This research investigates the capacity of an Applicant Tracking System (ATS) to enhance diversity and inclusivity in hiring practices, with a particular focus on Persons with Disabilities (PWD). The study presents an innovative adaptation of ATS whereby the threshold hire match is adjusted according to an individual's degree of ableism, as determined by their disability certificate. The findings illustrate that such an approach can significantly improve the inclusion of candidates with varying levels of disability, thereby promoting a more equitable recruitment process. The research corroborates previous studies asserting the benefits of workforce diversity and the efficacy of ATS in expediting recruitment procedures. The study underlines that swift responses to job applications can markedly enhance the employee lifecycle, particularly in small to medium-sized enterprises. Despite acknowledging the need for a cautious application of ATS to prevent unintended biases, especially in specific leadership roles, the research concludes that integrating diversity and inclusivity parameters into ATS could revolutionize the drive towards creating equitable workplaces and societies. The study stresses the need for further research to ensure broader applicability across various sectors, suggesting that these advancements in recruitment procedures could greatly improve workforce diversity and inclusivity.

Keywords: diversity, inclusion, application tracking system, hiring, disability

1.0 Introduction

In the last few decades, numerous organizations have shifted to e-recruiting models that screen candidate resumes for quality. As per Strohmeier (2010), e-portfolios are now frequently used in recruitment to find the ideal candidate matches. These effective models are predominantly employed by organizations with a workforce of over 500. Barber (1998) suggests that it is crucial for employers to sustain a consistent job application cycle. One technology that has surged in popularity is Applicant Tracking Systems (ATS). These systems are designed to optimize and streamline the recruitment process. Hom (2012) states that ATS can significantly reduce time and space dependencies for employers. ATS consists of two components, one designed for job seekers and the other for employers. The job seeker uploads their resume, cover letter, and other necessary documents along with their basic details. Concurrently, the employer outlines the job description, roles, responsibilities, and required skills. The ATS compares the applicant's data with the employer's criteria, producing a 'hire match'. Recruiters set a threshold

https://doi.org/10.52202/070182-0002

hire match for each role to determine the suitable candidate. If an application exceeds the threshold, the candidate is advanced to the next stage in the selection process. Otherwise, a rejection notice is sent. ATS is typically created using various programming languages, and this particular work focuses on the implementation using Python. However, given the high-risk measure, this technique is not typically applied when hiring for niche leadership roles or when the role or organization does not utilize ATS for recruitment. Therefore, this work's scope is confined to roles recruited via ATS, with an organization that values workforce diversity and inclusion.

In this context, the work at hand is centered on leveraging ATS to hire candidates with disabilities. The extent of disability can be determined by a government-issued disability certificate. Depending on the degree of disability, the threshold hire match can be adjusted. While this research is focused on applicants with disabilities, the same approach could be extended to other underrepresented candidate categories. As Lindsay et al. (2018) state, two-thirds of the disabled population is ready and willing to work, but they often struggle to find employment. This refinement in the recruitment tool might significantly expedite their job search, providing them with better employment opportunities.

2.0 Literature Review

Peichava (2022), in the paper "Data Analysis from the Applicant Tracking System," underlines that the conventional method of using Excel spreadsheets for hiring is considerably more time-consuming compared to an Applicant Tracking System (ATS). The ATS enhances the recruitment process by automating various tasks, such as screening resumes, tracking candidates, and scheduling interviews. This systematic organization and searchability of applicant data based on specific criteria significantly decreases the time and effort invested by hiring managers. Therefore, an ATS enhances hiring efficiency, allowing HR professionals to concentrate more on the strategic aspects of talent acquisition. Ashton & Morton (2005) further emphasize the importance of using an ATS to ensure the timely placement of leadership roles in human resources.

The paper "Role of Information Technology in Human Resource Management of SME: A Study on the Use of Applicant Tracking System" by Mukherjee et al. (2014) elaborates on the principles of e-recruitment. E-recruitment involves leveraging online platforms and technologies to attract, screen, and select job applicants. This process typically includes posting job vacancies on company websites, job boards, and social media, followed by the electronic submission of applications and resumes. E-recruitment expands an organization's reach to a larger candidate pool, accelerates application processing, and enhances communication with applicants.

For a diverse and inclusive workforce, organizations must proactively incorporate these principles into their recruitment strategies. E-recruitment platforms effectively enable this implementation. Organizations can create inclusive job descriptions and advertisements that emphasize equal opportunities, thereby inviting applicants from varied backgrounds. The use of clear and unbiased language can attract a more diverse range of applicants and help mitigate unconscious bias in the selection process.

https://doi.org/10.52202/070182-0002

E-recruitment systems can be designed to support blind recruitment techniques, too. By anonymizing applicant data, such as names, gender, and demographic details, biases can be eliminated during the initial screening stages. This approach ensures that candidates are assessed purely on their qualifications, skills, and experiences, leading to a fair and unbiased selection process. Moreover, e-recruitment platforms can use diverse sourcing channels. Leveraging social media networks, niche job boards, and professional forums, organizations can actively engage underrepresented groups. Promoting job opportunities via these channels can enhance visibility amongst diverse candidates and attract a more varied applicant pool. As Nachdeva (2021) noted, rapid advancements in technology now allow individuals with special needs to have equal access to their workplaces. Leumer further discusses how an ATS reduces the recruitment process costs, such as the need to hire personnel specifically for application screening.

In addition, as per AbilityNet (2021), to make content accessible to visually impaired individuals, several tools and devices are required that can convert text content into speech. The inclusion of text-to-speech functionalities in ATS, using Object Character Recognition (OCR) technology, can greatly enhance accessibility for visually impaired applicants. Antonova and Ivanova (2021) argue that with the help of advanced technologies like artificial intelligence (AI) and bots, HR professionals can refocus their efforts on more strategic roles in the organization. As a result, technology and AI are becoming integral to HRM operations.

3.0 Data Collection

Each job posting garners multiple resumes. Each of these resumes is tokenized, meaning it is broken down into individual words, which are then compared to the job description to form a matrix. This process is executed by the software during the parsing of the resume. After conducting cosine similarity and matrix operations, a unique result is obtained for each applicant within a specific role. This result is compared with the threshold hire match set by the employer. For minority candidates, a modified threshold hire match is applied based on the selection indicated by the applicant during the application process.

Peicheva (2022) emphasizes that this approach can significantly reduce the waiting time for applicants and also curtail recruiters' efforts to hire. Furthermore, it shortens the period between job posting and candidate hiring. From the employer's perspective, this method enhances the job offer acceptance rate. In this context, Eckhardt and Laumer's (2009) exploration of the architecture of the applicant tracking system is particularly relevant. This work places a strong emphasis on the inclusion of minority candidates in the hiring process.

Figure 1

Block Diagram of Application Tracking System

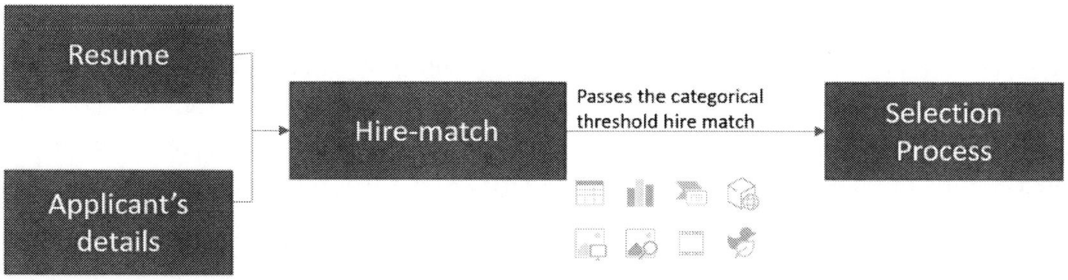

Let's consider a mathematical relationship between the expectations of an employer and a job applicant:

We'll define "JD" as a variable encompassing all words used by the employer in the job description, and "Resume" as a variable containing all words used by the job applicant in their resume.

[JD] represents a matrix containing all the words of the employer's criteria, while [Resume] is a matrix comprising all the words of the job applicant's resume.

The comparison between [JD] and [Resume] is done using cosine similarity. Cosine similarity is a measure of similarity between two non-zero vectors defined in an inner product space. It is calculated as the cosine of the angle between the vectors, which is equal to the dot product of the vectors divided by the product of their lengths. Cosine similarity is often used in data analysis, information retrieval, text mining, and data mining to compare documents, words, or other entities based on their features (Srivastava, 2023).

The 'Resultant' is calculated using the following formula for cosine similarity:

[Resultant] = Cos Θ = vec([JD]) . vec([Resume]) / (mag([JD]) * mag([Resume]))

Here, 'vec' denotes the vector of the matrix, in which the dot product is used, and 'mag' refers to the magnitude of the variables.

Following this, the Hire Match Fraction (HMF) is derived from the second element (index 1) of the 'Resultant' matrix.

Finally, the Hire Match (HM) is computed by multiplying the HMF by 100.

In this mathematical relationship, the 'Resultant' represents the cosine similarity between the job description and the resume.

https://doi.org/10.52202/070182-0002

4.0 Analysis

During the course of working with visually impaired students, the author developed an understanding of the extent of disability that exists among Persons with Disabilities (PWD) in his native country. This knowledge was further nuanced by the observation that each type of disability is categorized based on the percentage of ableism. The percentage of ableism for PWD candidates can be ascertained via the disability certificate. Multiple surveys helped shed light on various accessibility challenges faced by these individuals. This subsequently led to the conceptualization of reducing the thresholding to an ideal 50%. The study utilized a sample resume and a sample job description of a person with 100% ableism for examination.

In this research, candidates with 100% ableism were tested against a 60% threshold hire match. The program then inquires if the candidate identifies as having a disability. If affirmed, the percentage of disability is recorded. If the disability surpasses 50 percent, the threshold hire match is reduced to 50 percent. Conversely, if the level of disability is less than 50 percent, the threshold hire match is adjusted to 55 percent. This approach advocates for the inclusion of candidates with disabilities in the hiring process.

Bombaci & Pejchar (2022) propose that academic recruitments necessitate frameworks and tools for fostering diversity and inclusion in hiring practices. Some organizations, acknowledging the importance of diversity, equity, and inclusion in hiring, engage external consultants to conduct workshops, develop demographic assessments, and devise plans to implement DEI in their recruitment process to foster inclusivity (Gassam & Salter, 2020). For instance, the analysis of inclusive hiring for diversity in the healthcare industry, and the implementation of diversity hiring in the radiology department, provides demonstrative evidence of such efforts (Glastonbury et. al, 2021).

5.0 Findings & Conclusions

The findings of this research illustrate the ability of an Applicant Tracking System (ATS) to effectively screen resumes of people with disabilities and other minority groups, mirroring the studies conducted by Mukherjee et al. (2014) and others. The ATS can be integrated with considerations for disabilities, thereby enhancing the diversity and inclusivity of hiring processes. In addition, rapid responses to job applications streamline the recruitment process for both the job seeker and recruiter, thereby fostering inclusivity in employment. A critical finding from this study is that it is feasible to adjust the threshold hire match in the ATS based on an individual's percentage of ableism, determined by their disability certificate. This is consistent with the notion of fostering inclusivity among a diverse set of employees proposed by Caker et al. (2003). It was observed that when the disability exceeds 50 percent, reducing the threshold hire match to 50 percent potentially promotes the inclusion of candidates with higher levels of disability. Moreover, this research is in line with the work of Nawas (2019) and Fradella (2018) who highlighted the benefits of an ATS in preventing application duplication and meeting inclusion targets in university hiring processes, respectively. Yet, as Scepura (2020) cautions, the use of ATS in specific leadership roles could inadvertently introduce biases, suggesting that careful application of the system is necessary to avoid unintended consequences.

Echoing Lele's (2002) assertion, this research reaffirms the notion that hiring for diversity can result in higher-performing employees and a significant return on investment. In this regard, the application of an ATS for inclusive hiring represents a significant step towards achieving a diverse and inclusive workforce. There is ample evidence to conclude that this research underscores the transformative potential of integrating diversity and inclusivity parameters into the ATS. Despite the need for further research and development to ensure the wider applicability and scalability of such systems across different sectors, it's evident that this adaptive approach can contribute significantly towards creating more equitable workplaces and societies.

https://doi.org/10.52202/070182-0002

References

AbilityNet. (2021). *Vision impairment and computing*. AbilityNet. Retrieved July 1, 2023, from https://abilitynet.org.uk/factsheets/vision-impairment-and-computing

Antonova, K., & Ivanova, P. (2021). Robotization and artificial intelligence-A new model of cooperation and/or interaction between machines and people in the workplace. *HR and Technologies*, (1), 60-76.

Ashton, C., & Morton, L. (2005). Managing talent for competitive advantage: Taking a systemic approach to talent management. *Strategic HR review, 4*(5), 28-31. https://doi.org/10.1108/14754390580000819

Barber, A. E. (1998). *Recruiting employees: Individual and organizational perspectives*. Sage Publications.

Bombaci, S. P., & Pejchar, L. (2022). Advancing equity in faculty hiring with diversity statements. *BioScience, 72*(4), 365-371.

Cakar, F., Bititci, U. S., & MacBryde, J. (2003). A business process approach to human resource management. *Business Process Management Journal, 9*(2), 190-207. https://doi.org/10.1108/14637150310468

Eckhardt, A., & Laumer, S. (2009). An IT-architecture to align e-recruiting and retention processes. *International Journal of E-Services and Mobile Applications (IJESMA), 1*(2), 38-61.

Fradella, H. F. (2018). Supporting strategies for equity, diversity, and inclusion in higher education faculty hiring. In L. K. Fuller & L. K. Malaguti (Eds.), *Diversity and inclusion in higher education and societal contexts: International and interdisciplinary approaches* (pp. 119-151). Springer International Publishing. https://link.springer.com/chapter/10.1007/978-3-319-70175-2_7

Gassam, J. Z., & Salter, N. P. (2020). Considerations for hiring external consultants to deliver diversity trainings. *Consulting Psychology Journal: Practice and Research, 72*(4), 275.

Glastonbury, C. M., Bucknor, M., Wall, S. D., & Hess, C. P. (2021). Hiring through the lens of diversity: strategies to create diverse departments. *Academic Radiology, 28*(12), 1775-1778.

Holm, A. B. (2012). E-recruitment: towards an ubiquitous recruitment process and candidate relationship management. *German Journal of Human Resource Management, 26*(3), 241-259.

Laumer, S., Maier, C., & Eckhardt, A. (2015). The impact of business process management and applicant tracking systems on recruiting process performance: an empirical study. *Journal of Business Economics, 85*(4), 421-453.

Lindsay, S., Cagliostro, E., Albarico, M., Mortaji, N., & Karon, L. (2018). A systematic review of the benefits of hiring people with disabilities. *Journal of occupational rehabilitation, 28*(4), 634-655.

Ly-Le, T.M.(2022). Hiring for gender diversity in tech.*Journal of Management Development, 41*(6), 393-403.

Mukherjee, A. N., Bhattacharyya, S., & Bera, R. (2014). Role of information technology in human resource management of SME: A study on the use of applicant tracking system. *IBMRD's Journal of Management & Research, 3*(1), 1-22.

Nacheva, R. (2021). Digital workplace accessibility for visually impaired people. *HR and Technologies*, (2), 76-91. https://hrandtechnologies.com/wp-content/uploads/2021/07/HR-and-Technologies-2-2021.pdf

Nawaz, N. (2019). Artificial intelligence face recognition for applicant tracking system. *International Journal of Emerging Trends in Engineering Research, 7*(12). https://doi.org/10.2139/ssrn.3521931

Peicheva, M. (2022). Data analysis from the applicant tracking system. *Choveshki Resur. Tehnol. HR Technol. Creat. Space Assoc*, 2, 6-15. https://hrandtechnologies.com/wp-content/uploads/2021/07/HR-and-Technologies-2-2021.pdf

Scepura, R. C. (2020). The challenges with pre-employment testing and potential hiring bias. *Nurse Leader, 18*(2), 151-156.

Srivastava, R. P. (2023). A new measure of similarity in textual analysis: Vector similarity metric versus cosine similarity metric. *Journal of Emerging Technologies in Accounting*, 1–14. https://doi.org/10.2308/JETA-2021-043

Strohmeier, S. (2010). Electronic portfolios in recruiting? A conceptual analysis of usage. *Journal of Electronic Commerce Research, 11*(4), 268-280. http://web.csulb.edu/journals/jecr/issues/20104/Paper2.pdf

Acknowledgments

I would like to thank Professor J. Alexis, PMP, for his continuous guidance, during this work. I would like to extend my gratitude to the Northeastern University for allowing me to explore all of the available resources; I especially thank my parents and sister who continuously motivated me towards my education and endeavors.

 https://doi.org/10.52202/070182-0002

Psychological Safety – Agile Incorporating Diversity, Equity, and Inclusion Approaches

Dr. Jeffrey Thompson
College of Professional Studies
Northeastern University – Toronto
jeff.thompson@northeastern.edu

Abstract

This paper summarizes the feedback shared by attendees from a majority Canadian audience at a few presentations on Psychological Safety – Agile Incorporating Diversity, Equity, and Inclusion Approaches. During the presentation, the participants provided interactive input via Slido. Since the creation of the Agile Manifesto in February 2001, many organizations have sought to deliver value to their customers faster. The managers of these organizations realize that they need to remove bureaucracy and streamline these processes. On the human side, we must develop better tactics to be more inclusive and equitably as we gain an increasingly diverse workforce globally. We must create a more welcoming environment inside our Agile teams to ensure we meet our deliverables. To be more at ease, open, and productive in their team duties, team members must have a sense of belonging. By respecting individual diversity in our workspaces and effectively addressing our biases, we hope to foster extraordinary creativity and more effective delivery to customers. Adopting psychological safety in our workplaces may be the key to releasing creativity and producing excellent results within our Agile spaces.

Keywords: psychological safety, agile, diversity, equity, inclusion

1. Introduction

This article starts with the view that people can be most effective if they do not have concerns for their safety. A team environment that fosters psychological safety, racial and ethnic diversity, and team cohesion can reduce employees' tendency to be on guard and increase well-being, inclusion, and engagement at work (Brassel, Shaffer, & Travis, 2023). The Agile environment expects people to interact more compared to the waterfall project delivery method, particularly within self-governing teams (Mergel, Ganapati, & Whitford, 2021). The traditional management approach is command and control, while the Agile approach is leadership and collaboration (Zavyalova et al., 2020). Let us level-set and ensure everyone knows the difference between the Agile and Waterfall (or traditional) methods. The Agile Manifesto has been in existence since 2001, and it has four (4) values and 12 principles (Beck et al., 2001). Agile tends to be more flexible than Waterfall (Andrei, Casu-Pop, Gheorghe, & Boiangiu, 2019). Regarding control, the traditional approach is process-centric, while the Agile approach is people-centric (Zavyalova, Sokolov, & Lisovskaya, 2020). There tend to be inconsistent approaches to recording lessons learned (Rowe & Sikes, 2006). One approach is to record lessons learned after completing a project and do nothing with them (Rowe & Sikes, 2006). Alternatively, teams record lessons learned and examine them before beginning new initiatives (Rowe & Sikes,

https://doi.org/10.52202/070182-0003

2006). Sprint retrospective sessions happen more frequently (Ozoliņ, 2018). The intent is to adapt changes for the following sprint, and the event is typically held between sprints, for example, after the previous sprint has ended and before the next planning meeting (Ozoliņ 2018).

We also need to be mindful of the changing face of our workforce globally. Our teams become more diverse each day, and we need to appreciate how we can treat each other equitably and include them in relevant conversations (Statistics Canada, 2017). Diversity is one of the most crucial success factors for an agile team (Attar & Abdul-Kareem, 2020). Diversity benefits businesses with revenue generation and promoting a positive work environment (Chandana, Srija, & Ramu, 2022; Knight, Raines, Barton, & Rodriguez, n.d.). The idea that race, class, gender, sexual orientation, ethnicity, nation, ability, and age do not act as unitary, mutually exclusive entities but rather as reciprocally creating phenomena is known as intersectionality (Collins, 2015). Doing the right thing should always come first and be fundamentally ingrained in an organization's brand (Knight et al., n.d.). However, companies could exclude a sizeable portion of their client base if they do not determine how they work for everyone (Knight et al., n.d.). For example, Nike's collaboration with Colin Kaepernick illustrates how this may succeed and boost its revenue (Knight et al., n.d.). However, we must understand how we engage people to create an environment where they can do their best work. Psychological Safety is the conviction that speaking out with ideas, questions, worries, or blunders will not result in punishment or humiliation (Zaidi, 2022). Unconscious bias could result in losing clients and skilled workers if not adequately addressed within organizations (Suveren, 2022). Additionally, a lawsuit due to harassment may be pretty expensive for the business (Suveren, 2022). Agile leaders reject the concept of the chain of command, motivate others, demonstrate empathy for others, engage in shared management, constantly communicate, and reinforce the shared vision, and steadfastly pursue the advancement of their teams, organizations, and society (Attar & Abdul-Kareem, 2020).

2. Literature Review

Starting With the Agile Manifesto

The Agile Manifesto has been in existence since 2001, and it has four (4) values and 12 principles (Beck et al., 2001). The four (4) values are: "Individuals and interactions over processes and tools; Working software over comprehensive documentation; Customer collaboration over contract negotiation; Responding to change over following a plan (Beck et al., 2001)." Matyskevic and Simanaviciene (2020) investigated the applicability of the Agile Manifesto in non-software applications. The Agile methods differ from conventional ones in that they are grounded in the values and principles of the Agile Manifesto (Schön, Escalona, & Thomaschewski, 2015).

Agile versus Traditional (Waterfall) Methodology

The Agile culture turns the bureaucracy's traditional organizational ideas on its head as Agile respects individuals and teams (Mergel et al., 2021). By using agile concepts, several corporate settings have undergone a significant transformation (Mergel et al., 2021). The most notable change is the project teams and company personnel's mindset (Bagiu, Avasilcăi, & Rusu, 2022). From an implementation perspective, Pool, Poole, Upjohn, and Hernandez (2019) noted that the Agile project management framework provides the tools to innovate and adapt smoothly

https://doi.org/10.52202/070182-0003

and simultaneously. Hamad and Al Fayoumi (2018) shared that agile methodology turns the project triangle on its head. This shift implies fixed time and cost for Agile methods, and the defined scope varies for each iteration for the client's needs (Hamad & Al Fayoumi, 2018). Most Waterfall projects had small teams, less than ten members, and a duration of less than six months (Andrei et al., 2019). The general trend is that teams prefer Agile when constant feedback is necessary (Andrei et al., 2019). Cheema (2022) created a study to examine well-known project management programs like MS Project, JIRA, Trello, Asana, and Redmine against the significance of feature sets. The findings suggested that JIRA is currently the top project management tool for managing IT projects (Cheema, 2022). The result also provided statistical proof that choosing JIRA as the best alternative to manage IT projects will rise by 23% if senior management supports it (Cheema, 2022).

From Lessons Learned to Retrospective

We must learn from failed projects to prevent such events from happening again (Rowe & Sikes, 2006). We must maximize project success to apply beneficial procedures and methods for completing current and upcoming tasks (Rowe & Sikes, 2006). Rowe and Sikes (2006) observed that the team should continue to record lessons learned throughout the project. Ozoliņš (2018) identified that agile methods emphasize the necessity of teamwork and communication in successful product development. Scrum is one of the most popular agile methods involving several artifacts and practices, including conducting retrospectives (Ozoliņš, 2018). This activity serves the purpose of a team meeting to improve development processes after each sprint (Ozoliņš, 2018). It is a time-boxed event, planned and facilitated to obtain the best outcome, where team members analyze issues and design a strategy for future advancements (Ozoliņš, 2018).

Why is Diversity, Equity, and Inclusion Important

Grebić (2019) compared traditional project management methodology with agile methodology. Many organizations aim to provide an acceptable response to the rapidly changing corporate industry setting (Grebić, 2019). Cohen (2022) found that the workplace is a microcosm of society, exhibiting both development and political division on a general level. Since the shift to sustainability impacts company culture, the organization's management must carefully plan the migration to agile methodologies (Grebić, 2019). These include changes in the population caused by factors such as racial or ethnic identification, religion or spirituality, immigrant status, gender, and sexual orientation, as well as changes related to income, educational attainment, socioeconomic level, and literacy (Gill, McNally, & Berman, 2018). The most notable difference is the project teams and company personnel's mindset (Grebić, 2019). Gill et al. (2018) observed that Canada's workforce and demographics are evolving. Social movements in recent years have given employees more leverage to demand change, and they now want businesses to make measurable changes and remove their public neutrality (Cohen, 2022). Creating diverse recruiting metrics is a frequent initial step for organizations to improve DEI since it increases diversity representation and gives the impression that the organization is forward-thinking to the public (Cohen, 2022). However, there are still glaring gaps in effectiveness, mainly when keeping staff members from historically marginalized racial and ethnic groups (Cohen, 2022).

Tariman (2022) commented that no matter how difficult it may seem, racism still exists in countries like the US, and if we do not do anything about it, it will eventually catch up with

https://doi.org/10.52202/070182-0003

us. Strategies should emphasize "Diversity, Equity, and Inclusion" (DEI) rather than "Diversity" alone for transformation to be effective (Gill et al., 2018). They should also develop inclusive team environments to increase employee satisfaction and engagement (Gill et al., 2018). As the change begins, middle management must share accountability with senior leadership (Gill et al., 2018). Thompson and Thompson (2020) used multiple regression analysis to reveal greater racial diversity on larger boards, despite many businesses needing to disclose their diversity information. Unlike many other nations, Canada lacks information on racial and ethnic diversity (Thompson and Thompson, 2020). Countries like New Zealand use the collection of race and ethnicity data to create better regulations and safeguard particular communities (Thompson and Thompson, 2020). Without the relevant data, Canadian legislators cannot appropriately generate training or supportive legislation for visible minority workers (Thompson and Thompson, 2020).

Understand Our Unconscious Biases

Building an informal, authentic organizational culture is necessary for the effectiveness of administrative processes and frameworks that address diversity, equity, and inclusion in a society that appreciates, cherishes, and accepts all (Hirica, 2021). Hirica (2021) stated that "unconscious bias" refers to mental shortcuts that our brains use to categorize a scenario or a specific group of people to produce speedy reactions and cognitive efficacy. Everyone harbors unconscious prejudices influenced by their racial background, gender, age, experiences, and culture. These include proximity bias, confirmation bias, name bias, appearance bias, gender bias, and age bias (Suveren, 2022). Suveren (2022) identified that unconscious bias could harm interpersonal and group dynamics by favoring those from a similar social background, giving a particular gender the upper hand, or ignoring an alternative viewpoint put forth by someone with a different level of education. Unconscious bias suppression is a talent that requires practice and memory, just like any other ability (Suveren, 2022). People must recognize unconscious bias and unlearn bad habits (Meyer, 2018). People should be honest and honest about the topic and willing to adjust (Meyer, 2018).

Appreciating Intersectionality

Gopaldas (2013) stated that "intersectionality" describes how social identity systems like race, class, and gender influence various aspects of life, particularly those involving privilege and oppression. Statistics Canada (2017) projects that by 2036, the proportion of the visible minority population in Canada will be about 35%. Foss and Klein (2022) observed that one of the most significant developments of the past ten years has been the growth of woke corporations (and the broader phenomenon of woke capitalism), but more research is still needed. To fully comprehend the "woke" movement, they urged management scholars to apply theories from organizational design, strategy formulation, and organizational change and theories from organizational behaviour, human resource management, production and operations, nonmarket strategy, and other fields (Foss & Klein, 2022). The popular press, social media, and political institutions have been where most of the woke capitalism debate has occurred (Foss & Klein, 2022).

Benefits of Diversity

Peterson and Mann (2020) shared information on how the racial gap cost the US economy $1.6 Trillion over 20 years. They also shared solutions that the government, corporations, and individuals can implement to address the gap (Peterson & Mann, 2020). They

 https://doi.org/10.52202/070182-0003

encouraged individuals to educate themselves, advocate for themselves and embrace risk and delayed gratification to generate wealth (Peterson & Mann, 2020). Corporations would benefit from developing metrics to analyze, report, and react more effectively (Peterson & Mann, 2020). Larson (2017) shared the result of two years of research that found that diverse teams provide better decisions 87% of the time. These teams had people of varying ages, genders, and geographic origins (Larson, 2017).

Diversity, Equity, and Inclusion for Agile Teams

Cohen (2022) found that the workplace is a microcosm of society, exhibiting both development and political division on a general level. Knight et al. (n.d.) shared that scrum values include honesty, decency, and bravery. They believed in encouraging people to get to know one another (Knight et al., n.d.). The group's culture and ideals are crucial to establishing trust when discussing honestly (Knight et al., n.d.). One strategy is to develop additional mentorship programs that recognize and assist people who strongly emphasize cultural diversity, such as volunteering at a nonprofit or giving speeches informing minorities about potential future employment prospects at the company (Knight et al., n.d.). Hofstede Insights (n.d.) has a six-dimensional national culture model, measuring fundamental cultural values such as individualism vs. collectivism, power distance, and uncertainty avoidance. By analyzing these dimensions, Hofstede Insights (n.d.) helps clients understand the cultural differences that may affect their business operations and develop strategies to work effectively across cultures. Griggs and Louw (1995) shared that high-performance teams require talent, patience, and an awareness of diversity to manage effectively. Management should establish conditions that foster innovation, cooperation, and respect to obtain positive results with diverse teams (Griggs & Louw, 1995). They shared four stages to create a diverse work environment (Griggs & Louw, 1995). The first stage is making the relationship (Griggs & Louw, 1995). The second stage is disagreements arising (Griggs & Louw, 1995). The third stage entails team building, while the fourth entails capitalizing on differences (Griggs & Louw, 1995).

Chandana et al. stated that organizations must focus on ongoing improvements to stay ahead of the competition in this ever-evolving business climate. An agile workforce is required to force organizations to adapt to shifting market demands (Chandana et al., 2022). By preserving diversity in the workforce, organizations can increase their level of agility (Chandana et al., 2022). This diverse workforce aids businesses in their development across a range of industries (Chandana et al., 2022). Diversity is one of the most critical success factors for an agile team (Chandana et al., 2022). Creating diverse recruiting metrics is a frequent initial step for organizations to improve DEI since it increases diversity representation and gives the impression that the organization is forward-thinking to the public (Cohen, 2022). However, there are still glaring gaps in effectiveness, mainly when keeping staff members from historically marginalized racial and ethnic groups (Cohen, 2022).

Lean-Agile Leadership

Attar and Abdul-Kareem (2020) observed that the current business environment, characterized by increased complexity, ambiguity, uncertainty, and volatility, will be around for a while. Today's business world experiences constant upheaval and unheard-of change (Attar & Abdul-Kareem, 2020). Organizations must embrace and incorporate agility into their operations to address this situation (Attar & Abdul-Kareem, 2020). However, organizations must prioritize

https://doi.org/10.52202/070182-0003

developing agile leadership to achieve significant agility in line with the current hypercompetitive and volatile market (Attar & Abdul-Kareem, 2020). Agile leaders can establish guiding principles, strategies, and systems for seamless organizational agility (OA) shift (Attar & Abdul-Kareem, 2020). These leaders guide fostering a corporate culture that models and supports holistic agility (Attar & Abdul-Kareem, 2020).

Creating Psychological Safety

Zaidi (2022) defines psychological safety and recommends applying it throughout the engagement. The proper way to create a space where people can openly share their thoughts and ideas is in how the pre-event convening takes place (Zaidi, 2022). Zaidi (2022) discusses belonging and applying it in leadership and supervisory programs with a diversity and inclusion (D&I) component. Brassel et al. (2023) shared a report based on a survey of more than 3,000 employees across five countries who experiences emotional tax—specifically, being on guard to protect against bias due to race, ethnicity, and gender—are widespread in Australia, Brazil, Canada, the United Kingdom, and the United States, with 61% of employees from marginalized racial and ethnic groups indicating that they are on guard to bias and discrimination on their teams. Their findings suggest that a team environment that fosters psychological safety, racial and ethnic diversity, and team cohesion can reduce employees' tendency to be on guard and increase well-being, inclusion, and engagement at work (Brassel et al., 2023).

Thorpe-Moscon, Pollack, and Olu-Lafe (2019) found that over 21% of Canadians and nearly 22% of the labor force are people of color. If Canadian businesses are to prosper, they must become better at luring and nurturing this expanding workforce (Thorpe-Moscon et al., 2019). Canada had the fourth-worst employee retention rate in the world, with an average of 16%. Professionals identifying as Black, East Asian, or South Asian (33% to 50%) report being particularly vigilant about bias (Thorpe-Moscon et al., 2019). When faced with bigotry, 50% to 69% of Black, East Asian, and South Asian professionals genuinely intend to resign (Thorpe-Moscon et al., 2019). Emotional tax risks Health and well-being (Thorpe-Moscon et al., 2019). Professionals who are highly vigilant against bias include 22% to 42% of Black, East Asian, and South Asians who report having significant sleep issues (Thorpe-Moscon et al., 2019).

3. Methods

This paper uses secondary data to support its conceptual framework. It also used poll formation collected during presentations on the topic, 'Psychological Safety – Agile Incorporating Diversity, Equity and Inclusion Approaches.' using Slido (Slido.com, n.d.). The presentation took place over three sessions, with 62 persons attending and an overall engagement of 58%. The overall Q&A engagement was 13%. During the presentation, the participants responded to four (4) questions:

- How Familiar Are You with Agile?
- Do You Believe the Majority is Always Right?
- What Are Your Thoughts About Sharing Uncomfortable News?
- What Do You Think When You Hear DEI?

https://doi.org/10.52202/070182-0003

4. Results and Discussion

As Ringer (2007) shared, it is challenging for employees in a large complex organization to have most of the knowledge within single individuals. Therefore, the environment must foster effective collaboration so that the information is available where and when required (Ringer, 2007). Ciric et al. (2019) observed that agility is a hot topic. Agile emphasizes adaptability, acceptance of change, ongoing improvement, and robust engagement. There is a growing realization that agile has something to offer besides software development in the larger project landscape (Ciric et al., 2019).

As the author started the presentation, the initial approach was understanding where the audience was in their Agile knowledge. The Agile approach to project delivery differs from the traditional approach, and we should ensure that everyone is on the same page. Most attendees had limited exposure to Agile, with 16 persons, or 54% having less than one year of experience with Agile. See Figure 1 Response to the question related to familiarity with Agile. We can reject the null hypothesis even though only 48% of the participants answered the question.

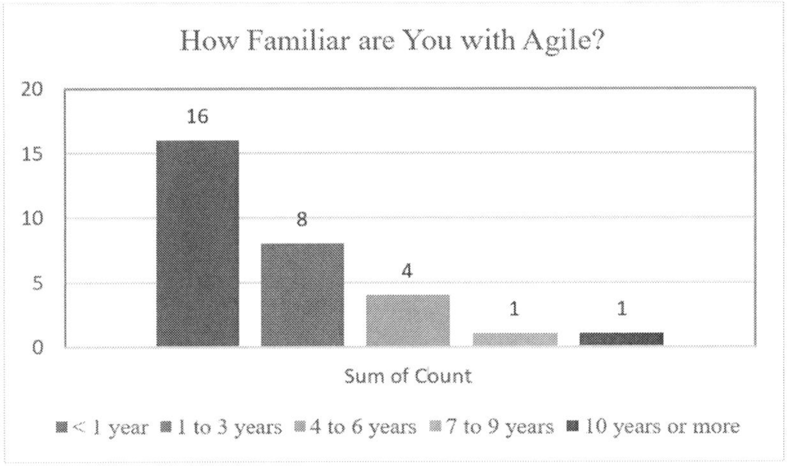

Figure 1 Response to the question related to familiarity with Agile

There were responses from 26 (42%) of the attendees to the question, 'Do You Believe the Majority is Always Right?' This result could mean that 36 (58%) participants were uncertain of their response. Figure 2 Response to the question related to the Majority Being Right. We can reject the null hypothesis even though only 48% of the participants answered the question. Based on the literature, the majority could have unconscious bias and make the wrong choices.

https://doi.org/10.52202/070182-0003

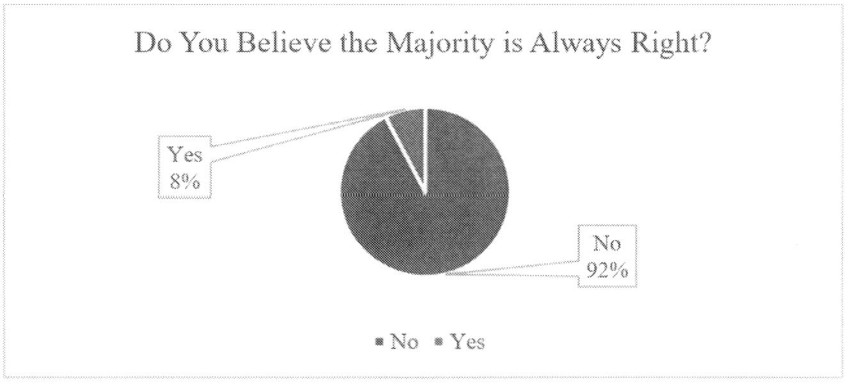

Figure 2 Response to the question related to the Majority Being Right

The remaining questions gauged the participants' views on the topics via Word Cloud analysis (Atenstaedt, 2012). Based on the responses to the following two questions, the author used Word Cloud Generator (n.d.) to generate the word clouds in Figure 3. Response to the question related to Thoughts About Sharing Uncomfortable News. There were responses from 37 (60%) attendees to the question, 'What Are Your Thoughts About Sharing Uncomfortable News?' This result could mean that 25 (40%) persons were uncertain of their response. The common theme was 'necessary' for 'people' to 'always' be 'transparent' and 'honest.' Based on the literature, people can only achieve this level of transparency when they have psychological safety.

Figure 3 What Are Your Thoughts About Sharing Uncomfortable News?

There were responses from 27 (44%) of the attendees to the question, 'What Do You Think When You Hear DEI? ' This result could mean that 35 (56%) persons were uncertain of their response. See Figure 4 Response to the question related to the thoughts on DEI. The message could read that we 'need' to establish a 'framework' to 'learn' to have 'more' 'acceptance,' 'diversity' and 'inclusion' with 'DEI.'

https://doi.org/10.52202/070182-0003

Figure 4 Response to the question related to the thoughts on DEI

5. Conclusion

Even though many people do not seem to understand Agile, it is a critical methodology. The likelihood of project success decreases without iterative planning, self-managing teams, and numerous releases of useful products provided early and with improved quality (Pool et al., 2019). Those companies serious about adopting Agile methodology should consider the appropriate HRM (Human Resource Management) structure to support the project management methodology (Zavyalova et al., 2020). The future of work is diverse, hybrid teams (Brassel et al., 2023). By forging connections, showing up, and holding themselves accountable for change, executives, team leaders, and individuals can take steps to foster inclusive team environments (Brassel et al., 2023). People are made stronger by appreciative inquiry and the chance to learn from those with various backgrounds and viewpoints (Knight et al., n.d.). Inclusive workplaces may reduce the emotional tax (Travis, Thorpe-Moscon, & McCluney, 2016). Employees may connect their personal and professional life in the best way possible, decreasing stress and increasing productivity (Travis et al., 2016).

References

Andrei, B. A., Casu-Pop, A. C., Gheorghe, S. C., & Boiangiu, C. A. (2019). A study on using waterfall and agile methods in software project management. *Journal of Information Systems & Operations Management*, 125-135. Retrieved from https://www.proquest.com/docview/2237828314

Ashikali, T., & Groeneveld, S. (2013). Diversity management in public organizations and its effect on employees' affective commitment. *Review of Public Personnel Administration, 35*(2), 146–168. https://doi.org/10.1177/0734371x13511088

Atenstaedt, R. (2012). Word cloud analysis of the BJGP. *British Journal of General Practice, 62*(596), 148. https://doi.org/10.3399/bjgp12x630142

Attar, M., & Abdul-Kareem, A. (2020). *The role of agile leadership in organisational agility. In Agile Business Leadership Methods for Industry 4.0.* Emerald Publishing Limited. https://doi.org/10.1108/978-1-80043-380-920201011

Bagiu, N., Avasilcăi, S., & Rusu, B. (2022). Traditional vs. agile: The challenge mindset shift in project management. *Review of Management & Economic Engineering, 21*(4), 301–311. https://doi.org/10.1016/j.promfg.2020.01.314

Beck, K., et al. (2001) *The agile manifesto. Agile Alliance.* Retrieved from http://agilemanifesto.org/

Brassel, S., Shaffer, E., & Travis, D. (2023, February 7). *Emotional tax and work teams: A view from 5 countries (Report)*. Catalyst. Retrieved from https://www.catalyst.org/reports/emotional-tax-teams/

Chandana, K., Srija, V., & Ramu, G. (2022, September). Work force diversity and agility: Diversity is a key factor to be agile. *International Research Journal of Modernization in Engineering Technology and Science, 4*(9) https://doi.org/10.56726/irjmets29674

Cheema, M. U. (2022). The choice of project management software by project managers; with the moderating impact of top management support: Project management softwares. *Journal of Innovation and Emerging Technologies, 2*(1), 39-63. Retrieved from https://readersinsight.net/jiet/article/view/2490

Ciric, D., Lalic, B., Gracanin, D., Tasic, N., Delić, M., & Medic, N. (2019b). Agile vs. traditional approach in project management: Strategies, challenges and reasons to introduce agile. *Procedia Manufacturing,* 39, 1407–1414. https://doi.org/10.1016/j.promfg.2020.01.314

Cohen, H. R. (2022). *Beyond the metrics: Understanding the importance of belonging in creating more equitable workplaces* (3080693) [Doctoral dissertation, University of Southern California]. ProQuest Dissertations and Theses Global.

Collins, P.H. (2015). Intersectionality's definitional dilemmas. *Annual Review of Sociology, 41*(1),1–20. https://doi.org/10. 1146/ annurev-soc-073014-112142

Foss, N.,& Klein, P.(2022).Why do companies go woke? *Academy of Management Perspectives, 1*. https://doi.org/10. 5465/ amp. 2021. 0201

Gill,G.K., McNally, M.J.,& Berman,V.(2018). Effective diversity, equity, and inclusion practices. *Healthcare Management Forum, 31*(5),196–199. https://doi.org/10. 1177/ 0840470418773785

https://doi.org/10.52202/070182-0003

Gopaldas, A. (2013). Intersectionality 101. *Journal of Public Policy & Marketing,* 32(1), 90–94. https://doi.org/10.1509/jppm.12.044

Government of Canada, Statistics Canada. (2017, October 25). *Number and proportion of visible minority population in Canada, 1981 to 2036*. Retrieved from https://www.statcan.gc.ca/en/dai/btd/othervisuals/other010

Grebić, B. (2019). Traditional vs agile project management in the service sector. *European Project Management Journal, 9*(2), 55–63. https://doi.org/10.18485/epmj.2019.9.2.6

Griggs, L. B., & Louw, L.-L. (1995). Diverse teams: Breakdown or breakthrough? *Training & Development, 49*(10), 45-50. https://www.proquest.com/docview/227000404

Hamad, R. M. H., & Al Fayoumi, M. (2018, November). Scalable agile transformation process (SATP) to convert waterfall project management office into agile project management office. In 2018 *International Arab Conference on Information Technology* (ACIT) (pp. 1-9). IEEE. https://doi.org/10.1109/ACIT.2018.8672701

Hirica, A. (2021). Unlocking Unconscious Biases into Impactful Design Work. *Design Management Review, 32*(3), 24–27. https://doi.org/10.1111/drev.12268

Knight, M., Raines, B., Barton, B., & Rodriguez, S. (n.d.). *Agilists discuss allyship*. (n.d.). Scrum Alliance. Retrieved March 29, 2023, from https://resources.scrumalliance.org/Article/agilists-discuss-allyshipm

Larson, E. (2017, September 17). Infographic: Diversity + inclusion = Better decision making at work. Cloverpop. Retrieved from https://www.cloverpop.com/blog/infographic-diversity-inclusion-better-decision-making-at-work

Matyskevic, J. & Simanaviciene, Z. (2020). Agile in commercial setup -- Mission impossible? *Public Security & Public Order / Visuomenes Saugumas Ir Viesoji Tvarka, 25*, 246–259. https://doi.org/10.13165/PSPO-20-25-24

Mergel, I., Ganapati, S., & Whitford, A. B. (2021). Agile: A New Way of Governing. *Public Administration Review, 81*(1), 161–165. https://doi.org/10.1111/puar.13202

Meyer, C. (2018). How to counteract unconscious biases. *Journal of Accountancy, 225*(2), 40. https://www.journalofaccountancy.com/issues/2018/feb/unconscious-bias.html

Ozoliņš, P. (2018). Preparation and facilitation of retrospective meeting in scrum process. *Information Technology and Management Science, 21*, 60–63. https://doi.org/10.7250/itms-2018-0009

Peterson, D. M., & Mann, C. L.(2020, September 1). *Closing the racial inequality gaps: The economic cost of Black inequality in the US.* Citi Global Perspectives & Solutions. Retrieved from https://ir.citi.com/%2FPRxPvgNWu319AU1ajGf%2BsKbjJjBJSaTOSdw2DF4xynPwFB8a2jV1FaA3Idy7vY59bOtN2lxVQM%3D

Pool, E. T. (2019, March 1). 'Agile' helping Mayo improve its projects. *Physician Leadership Journal, 6*(2), 28-33. https://www.proquest.com/docview/2186245610

Prestia, A. S. (2019). Sabotaging success: The role of unconscious bias. *Nurse Leader, 17*(6), 561-564. https://doi.org/10.1016/j.mnl.2019.02.005

Ringer, T. M. (2007). Leadership for collective thinking in the workplace. *Team Performance Management, 13*(3/4), 130–144. https://doi.org/10.1108/13527590710759874

Rowe, S. F. & Sikes, S. (2006). *Lessons learned: taking it to the next level*. Paper presented at PMI® Global Congress 2006—North America, Seattle, WA. Newtown Square, PA: Project Management Institute. Retrieved from https://www.pmi.org/learning/library/lessons-learned-next-level-communicating-7991

Schön, E., Escalona, M. J., & Thomaschewski, J. (2015). Agile values and their implementation in practice. *International Journal of Interactive Multimedia and Artificial Intelligence, 3*(5), 61. https://doi.org/10.9781/ijimai.2015.358

Slido. (n.d.). Audience interaction made easy. https://www.slido.com/

Suveren, Y. (2022). Unconscious bias: Definition and Significance. *Current Approaches in Psychiatry / Psikiyatride Guncel Yaklasimlar, 14*(3), 414–426. https://doi.org/10.18863/pgy.1026607

Tariman, J. D. (2022). Overcoming racism: A call to action for our collective and renewed commitment to building an ethos of diversity, equity, and inclusion (DEI). *Journal of Nursing Practice Applications & Reviews of Research, 12*(2), 1. https://doi.org/10.13178/jnparr.2022.12.02.1201

Thompson, J., & Thompson, R. F. (2020). Racial diversity in publicly traded companies. *International Journal of Sustainable Entrepreneurship and Corporate Social Responsibility, 5*(2), 66–85. https://doi.org/10.4018/ijsecsr.2020070105

Thorpe-Moscon, J., Pollack, A., & Olu-Lafe, O. (2019). *Empowering workplaces combat emotional tax for people of colour in Canada*. Catalyst. https://www.catalyst.org/research/emotional-tax-canada/

Travis, D., Thorpe-Moscon, J., & McCluney, C. (2016). *Emotional tax: How Black women and men pay more at work and how leaders can take action*. Catalyst. https://www.catalyst.org/research/emotional-tax-how-black-women-and-men-pay-more-at-work-and-how-leaders-can-take-action/

Zaidi, S. (2022). Integrating psychological safety: Using the all-in method to cultivate belonging

 https://doi.org/10.52202/070182-0003

and understanding in diversity and inclusion. *AI Practitioner, 24*(4), 37–42. https://doi.org/10.12781/978-1-907549-53-3-7

Zavyalova, E., Sokolov, D., & Lisovskaya, A. (2020). Agile vs traditional project management approaches: Comparing human resource management architectures. *International Journal of Organizational Analysis, 28*(5), 1095-1112. https://doi.org/10.1108/IJOA-08-2019-1857

Biography

Jeffrey is an Agile Release Train Engineer (RTE) in CORE Platform – Agile Practice at TD Bank. In his role at the bank, he is a project professional who acts like a chief Scrum Master and servant leader. He supports his teams and other stakeholders to deliver on their Objectives and Key Results (OKR) effectively. Jeffrey does what is necessary to take his teams to a marvelous WOW!

He has over 30 years of experience and proven effectiveness with successful implementations in the financial, pharmaceutical, brewing, Consumer Packaged Goods (CPG) industries and government (Ontario Lottery and Gaming). Jeffrey is a Past President of the Project Management Institute (PMI) Toronto chapter. He also gives back to the community as a York Region Alliance of African Canadian Communities (YRAACC) board member and volunteers in his church, including with the Knights of Columbus. He is a Black Employee Network (BEN) member at TD Bank. Jeffrey has a Doctor of Business Administration (DBA) in General Business from Northcentral University (NCU), Arizona, his MSc. in Computer Information Systems from the University of Phoenix, and his BSc. in Electrical and Computer Engineering from the University of the West Indies (UWI), St Augustine, in Trinidad. He is a researcher, and his ORCID number is https://orcid.org/my-orcid?orcid=0000-0002-8776-7719.

Jeffrey is an accomplished and dynamic Certified Project Manager (PMP), Six Sigma Green Belt, MS Project Trainer, Agile, Scaled Agile Framework (SAFe), and ITIL professional. He also mentors professionals and teaches project management.

Acknowledgments

- Thank you, Jacques Alexis, and John Bleakney, for inviting me to present my topic at your Roundtable: Evidence-Based Project Management event.

Rising Above Waste: Leading the Way in Waste Management for the Construction Industry

S. Tahereh Ahadisarkani [1*], Sheida Khandadashpoorghomi[2], Ava Ghased[3], Alireza Shahparvari [4], Mohammad Abniki[5], Adel A. Zadeh [6]

[1*]College of Professional Studies, Northeastern University, Toronto, Canada; ahadisarkani.s@northeastern.edu
[2]College of Professional Studies, Northeastern University, Toronto, Canada; sheidakhandadashpo.s@northeastern.edu
[3]College of Professional Studies, Northeastern University, Toronto, Canada; ghased.a@northeastern.edu
[4]College of Professional Studies, Northeastern University, Toronto, Canada; shahparvari.a@northeastern.edu
[5]College of Professional Studies, Northeastern University, Toronto, Canada; abniki.m@northeastern.edu
[6] College of Professional Studies, Northeastern University, Toronto, Canada; a.zadeh@northeastern.edu.edu
* Correspondence:ahadisarkani.s@northeastern.edu

Abstract

The Canadian construction industry significantly contributes to the country's waste production. Despite initiatives advocating for sustainability, issues surrounding construction material waste persist. This systematic review, utilizing the PRISMA normative method, sought to understand the influence of government regulations and pinpoint the challenges in executing effective waste management strategies in the Canadian construction domain. Out of 580 titles and abstracts, 71 full papers were extensively analyzed to assess regulations, policies, and initiatives aimed at diminishing construction material waste. The findings highlight that the primary barriers include insufficient regulations, suboptimal implementation strategies, and a lack of focused leadership on construction material waste management. These issues are further intensified by an absence of sustainable practices, poor regulatory enforcement, disjointed organizational coordination, and the inherent complexity of waste management systems. Addressing these barriers is pivotal for promoting efficient waste management principles in the Canadian construction industry, paving the way for a sustainable future.

Keywords: waste management, barriers, regulation, policy, leadership

1 Introduction

Construction waste in Canada includes various materials such as metals, wood, plasterboard, concrete, and others (Llatas, 2011). The construction industry in Canada generates approximately 9 Million tons of construction and demolition waste (CDW) annually, with wood products, asphalt, drywall, concrete, and masonry being the main components (Ahmed & Sobuz, 2019; Mahpour, 2018; Yeheyis et al., 2013a). Construction waste accounts for about 27% of the total municipal solid waste in Canada (Asgari et al., 2017).

Efforts are being made to divert construction waste from landfill and promote recycling and reuse. However, there are still challenges in effectively managing construction waste. The industry contributes the largest proportion of waste to landfill in Canada, and similar patterns exist in other large economies such as the UK, US, Brazil, Australia, and Hong Kong (Ajayi et al., 2017). Construction and demolition (C&D) debris is a growing concern for cities worldwide due to its increasing generation and disposal in landfills (Abarca-Guerrero et al., 2017; Swetha et al., 2022).

 https://doi.org/10.52202/070182-0004

The negative environmental impacts of waste highlight the importance of reducing construction waste (Ajayi et al., 2015; Amuna et al., 2021). Construction waste management (CWM) regulations have significantly enhanced on-site CDW in Canada (Earle, 2021).
As such, there has been a growing interest in sustainability within the industry, focusing on reducing the environmental impacts associated with construction activities (Al-Otaibi et al., 2022; Andersson & Buser, 2022; Ortiz et al., 2009). Considering that, Canadian governments have implemented several successful programs for managing construction and demolition waste, including the Ontario Ministry of the Environment's Construction Site Waste Management program and the City of Vancouver's Green Demolition Bylaw. By reducing environmental impacts and promoting sustainability, these initiatives illustrate the potential of effective waste management practices (Yeheyis et al., 2013a). For example, the Green Building Policy for Municipal Buildings in Vancouver requires that new municipal buildings incorporate a minimum of 40% reused or recycled materials. This policy has helped promote sustainable construction practices and reduce waste sent to landfills. Similarly, the Greater Toronto Area Construction Waste Coalition has developed guidelines for construction waste management that emphasize waste reduction, Reuse, and recycling, demonstrating the potential benefits of waste management reuse in the construction industry.

However, there is still a need for improvement in waste management practices and the adoption of sustainable strategies. Inadequate regulations and implementation strategies that focus on the management of construction material waste are the main causes of inadequate waste management in Canada (Earle, 2021; Mendis et al., 2015; Yeheyis et al., 2013a). This problem is compounded by a lack of sustainable practices, lack of enforcement of regulations, lack of coordination among organizations, and the complexity of waste management systems. The implementation of sustainable construction material waste management is essential to reducing the environmental impacts associated with improper disposal of construction material waste.

This paper conducts a systematic literature review using the PRISMA approach to explore regulations, barriers, and strategies in construction material waste management in Canada. The research objectives include identifying and analyzing waste management strategies, barriers, and regulations in the construction industry. It aims to propose solutions and strategies to overcome implementation barriers and contribute to knowledge in this field. The findings can inform policymakers, industry professionals, and researchers, promoting sustainable waste management practices in construction. Ultimately, the paper seeks to provide a comprehensive understanding and promote efficient waste management in the Canadian construction industry.

2 Review methodology

To examine the current regulations, policies, and initiatives that are targeted at reducing construction material waste in Canada, this systematic literature review will utilize the PRISMA (Preferred Reporting Items for Systematic Reviews and Meta-Analyses) methodology. The PRISMA method consists of 27 subitems that encompass the various stages of a systematic review, such as defining eligibility criteria, identifying information sources, conducting search strategies, selecting relevant studies, extracting outcomes, and synthesizing data (Moher et al., 2015). By adhering to the PRISMA guidelines, we ensured transparency and rigor in our review process.

https://doi.org/10.52202/070182-0004

Our review was conducted in three phases: planning, development, and analysis. During the planning phase, we outlined the research objectives, refined the research question, and established the research strategy and protocol. In this stage, Microsoft Teams was selected as our communication platform to share files, synchronize modifications, and share updates. In addition, we discussed and identified the list of keywords to include in the search strings. Searches were performed by employing combinations of key terms and using the Boolean operators AND and OR : ("Waste" OR "waste manage*") AND ("Construction" OR "Building Construction") AND ("Barrier*" OR "Obstacle*" OR "Impediment*" OR "Hurdle*", "Restriction*s" OR "Constraints *" Or "Challenge*" OR "Limit*).

In the development phase, we used the PRISMA flow chart, shown in Figure 1, to identify and include relevant studies from selected databases. A comprehensive database search was carried out using keywords chosen over several meetings. In the identification stage, bibliographical searches were conducted across four comprehensive databases: Web of Science (n = 685), Google Scholar (n =155), Science direct (n=26), and Advanced Science letter (n=2). The search results were limited to article papers and review papers. The option search in Web of Science, Google Scholar, Science direct, and Advanced Science letter was "title", "abstract", and "keyword".
In stage two of the review process, the authors screened the retrieved journal papers (n=868) by removing duplicates in Zotero. This resulted in 240 papers. Following the PRISMA methodology the team evaluated the titles and abstracts of the works for potential inclusion in the following stage. In the next stage of selecting and including papers in the review, and setting the parameters for the scope of the review, some eligibility criteria were developed and applied to the data collected during previous stages: The inclusion criteria were: 1) peer-reviewed journals, 2) must be related to the construction industry 3) The papers should discuss and identify barriers and challenges in adopting or implementing, 4) The papers examine barriers either qualitatively or quantitatively or in review papers. The Exclusion criterion was: The papers did not mention barriers, techniques, challenges, and policies of construction material waste and did not explore those in-depth. In the initial review team screening, 161 works were included, of which the ultimate inclusion of 72 of these works was doubtful.

Finally, in the analysis phase, the authors thoroughly assessed the selected papers, drawing insights on barriers and challenges, and deriving conclusions and recommendations. The authors read and analyzed all included papers separately and combined the acquired data in a shared MS Excel file. When they could not agree on data, they discussed and collaborated to reach a common view. Consequently, qualitative data were extracted from 72 included papers. The review team comprehensively collected data relevant to the publication of the articles in an Excel spreadsheet, including information such as author names, publication year, journal name, study description, and types of projects.

 https://doi.org/10.52202/070182-0004

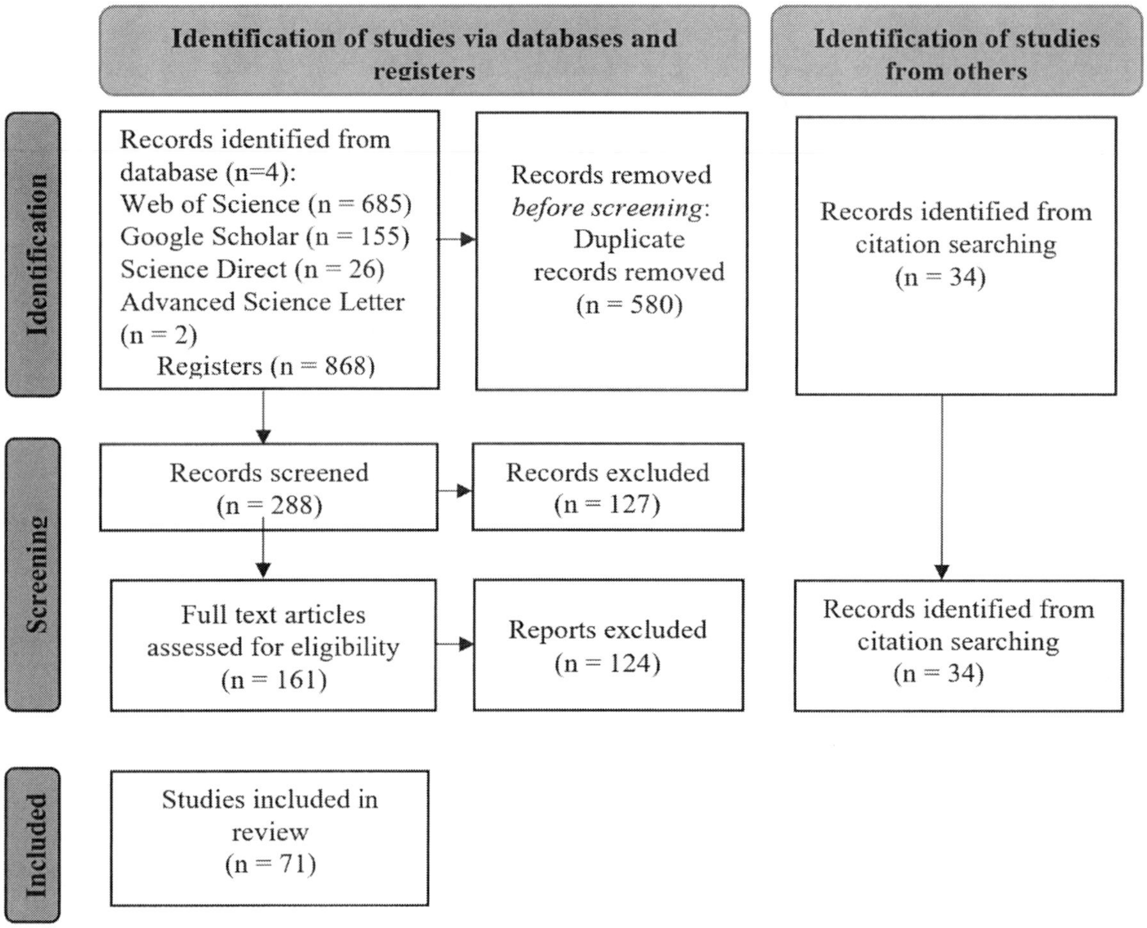

Figure. 1. Flowchart of Process for selecting studies

3 Results and Discussion

Several studies have been conducted to identify the barriers to sustainable waste management in the construction industry and propose potential solutions. This article aimed to integrate and expand on some of these studies to provide a comprehensive overview of the barriers and optimal solutions to sustainable waste management in the construction industry.

3.1 Waste management Strategies

The construction industry employs various waste management strategies to address the challenges associated with construction waste, including long-term policies, stakeholder commitment, site management practices, cultural change, efficient logistic and procurement processes, and on-site recycling.

As such efforts, Tam (2009) underscores the necessity of long-term policies and strategies aimed at coordinating construction stakeholders in the implementation of waste management practices, despite the substantial investment required for the development of facilities and acquisition of

equipment. This perspective emphasizes the importance of fostering collaboration and coordination among various stakeholders to effectively address construction waste management challenges.
On the other hand, Manowong (2012) draws attention to the significance of stakeholder commitment and the achievement of construction waste management goals, particularly in developing countries. This viewpoint highlights the need for active engagement and participation from stakeholders to ensure the successful implementation of waste management initiatives within the construction industry. Kulatunga et al. (2006) delve into the influence of attitudes and perceptions of the construction workforce on waste management strategies. Their work underscores the relevance of understanding the perspectives and behaviors of individuals involved in construction activities to devise effective waste management approaches. Ajayi et al. (2017) identify key site management practices that contribute to effective waste minimization, such as strict adherence to project drawings and the provision of waste skips. Their findings emphasize the significance of implementing sound management practices to mitigate waste generation during construction processes. In highly urbanized cities, Yu et al. (2021) recommend the adoption of effective practices, including circular economy approaches and green rating systems, to enhance construction waste management. This viewpoint highlights the need for innovative and sustainable approaches to waste management in densely populated urban areas. Cultural factors contributing to waste intensiveness in the UK construction industry are examined by Ajayi et al. (2016), who propose changes to reduce waste. This perspective highlights the role of cultural factors and the need for cultural shifts within the industry to promote waste reduction and more sustainable practices.

In addition to the main waste management strategies in the construction industry, there is a growing emphasis on the principles of the 3 Rs: Reduce, Reuse, and Recycle. These principles aim to minimize waste generation, maximize resource utilization, and promote sustainable practices throughout the construction lifecycle. The first principle, Reduce, focuses on minimizing waste generation at the source. This can be achieved through careful planning and designing, optimizing material use, and implementing efficient construction practices (Bao et al., 2020a; Bao & Lu, 2021; Begum et al., 2007; Mohammed et al., 2021; Ogunbiyi et al., 2014; Tam et al., 2005). Studies highlight the importance of waste reduction through lean construction techniques, such as just-in-time delivery and prefabrication, which can minimize material waste and improve overall project efficiency (Hao et al., 2022; Abarca-Guerrero et al., 2022). They also emphasize the significance of waste reduction strategies in the construction industry, including the use of modular construction and off-site fabrication, which can reduce material waste and improve resource efficiency (Bossink & Brouwers, 1996; Bui et al., 2020; Mostafa et al., 2016). The second principle, Reuse, involves finding opportunities to repurpose materials and components to extend their lifespan. This can be achieved through salvaging and reusing materials from demolition or renovation projects, as well as implementing strategies for the reuse of construction waste within the same project or in other construction projects (da Rocha & Sattler, 2009; Guerra et al., 2020; Junaid et al., 2022; Park & Tucker, 2017). Studies discuss the potential for reusing construction and demolition waste materials, such as concrete and timber, in new construction projects (Tam, 2009). They also highlight the importance of developing a systematic approach to identify and evaluate opportunities for material reuse in the construction industry, which can contribute to waste reduction and resource conservation. The third principle, Recycle, focuses on the proper management and recycling of construction waste materials. Recycling involves processing waste

 https://doi.org/10.52202/070182-0004

materials into new products or materials that can be used in construction or other industries (Bao et al., 2020b; Bao & Lu, 2021; Earle, 2021). Studies emphasize the importance of implementing effective recycling systems and infrastructure to divert construction waste from landfills (Ding et al., 2018; Duan et al., 2019; Ferronato & Torretta, 2019; Ortiz et al., 2009). They discuss the potential for recycling construction and demolition waste, such as concrete, asphalt, and metals, and highlight the environmental and economic benefits of recycling in the construction industry. By incorporating the principles of the 3 Rs into waste management strategies, the construction industry can significantly reduce waste generation, conserve resources, and promote sustainable practices (El-Haggar, 2007; Elshehawy et al., 2022; Foo et al., 2013). These principles align with the broader goals of achieving a circular economy, where materials are kept in use for as long as possible through reduction, reuse, and recycling, ultimately minimizing the environmental impact of construction activities (Hentges et al., 2022; Jahan et al., 2022; Low et al., 2020; Osobajo et al., 2022; Palafox-Alcantar et al., 2021).

3.2 Barriers

Several studies have identified various barriers that impede the implementation of waste management strategies in the construction industry. Lack of regulations, standards and incentives, limited stakeholder awareness, , cost-related concerns, insufficient infrastructure for waste management, and inadequate collaboration among stakeholders were reported impeding the effective management of construction waste management (Ajayi et al., 2015).

One significant barrier is the lack of regulations and standards specific to waste management in the construction industry. The absence of clear guidelines and requirements can lead to inconsistent practices and a lack of accountability (Kulatunga et al., 2006). Without clear requirements, there is a diminished sense of accountability, as there are no established benchmarks or metrics against which parties can be held responsible for their waste management actions. Furthermore, inconsistent regulations across different municipalities can make implementing consistent waste management practices difficult (Earle, 2021; Yeheyis et al., 2013b). The lack of a unified approach to (CDW) can give rise to confusion among businesses and organizations that operate in different areas that may require following other guidelines in each location, making it difficult and time-consuming to implement consistent waste management practices (Shooshtarian et al., 2022). This highlights the challenges and complexities in achieving consistent waste management practices in the construction industry. Besides, discrepancies in regulations may diverge based on the category of waste generated and the commercial sector, thereby augmenting the intricacy of waste management. In developing countries, waste mismanagement is a global issue, with open dumping and burning being the main waste treatment and disposal systems, leading to environmental contamination and social issues (Ferronato & Torretta, 2019). In Malaysia, the rapid growth of the construction sector has led to the need for careful review of existing environmental management strategies for construction waste, including policies, laws, and regulations, to address environmental challenges (Chang & Kumar, 2021). Municipal corporation may mandate commercial establishments to classify their waste into distinct categories, such as hazardous waste, organic waste, and recyclables, before disposing of them. On the other hand, a dissimilar municipal corporation may not have instituted any policies concerning waste segregation, thereby causing an absence of uniformity in C&D disposals (Shooshtarian et al., 2022).

Technical hurdles and a lack of expertise and knowledge in waste management can also hinder the implementation of sustainable waste management strategies (Ghafourian et al., 2021; Tsai et al., 2020). The construction industry may face difficulties in adopting modern technologies and practices due to a lack of awareness and understanding (Saad et al., 2022). Furthermore, insufficient supervision and weak policies have been identified as major obstacles to achieving sustainability in the construction industry (Ghafourian et al., 2021)
The role of leadership and management is crucial in overcoming these implementation barriers. Strong leadership commitment and top management support are essential for driving change and promoting sustainable waste management practices (Kulatunga et al., 2006). Leaders can set the tone for waste reduction and recycling efforts, establish waste management plans, and allocate resources for waste management initiatives (Kulatunga et al., 2006). Effective communication and collaboration among stakeholders, including contractors, developers, and government agencies, are also vital for successful waste management implementation (Saad et al., 2022)
The importance of stakeholder awareness and cooperation in promoting sustainable construction and demolition (C&D) waste management is supported by several studies. Jin et al. (2019) conducted a science mapping approach to evaluate C&D waste management research and emphasized the need for research and collaboration among stakeholders to address challenges and gaps in the field. Shooshtarian et al. (2020) focused on market development for C&D waste materials in Australia and found that stakeholder cooperation is crucial for establishing end markets for recyclable materials. Wu et al. (2020) discussed the case study of Hong Kong, highlighting the role of stakeholders in driving effective C&D waste management for sustainable development. Shooshtarian et al. (2022) explored the challenges and opportunities in the Australian C&D waste management system, emphasizing the importance of stakeholder involvement in improving the system. Palafox-Alcantar et al., (2021) surveyed stakeholders involved in waste management in Birmingham, UK, and emphasized the significance of stakeholder cooperation in facilitating the transition to a circular economy. Overall, these studies underscore the integral role of leadership and stakeholder awareness and cooperation in promoting sustainable C&D waste management practices.

The lack of space can pose challenges in implementing consistent construction waste management practices, as evidenced by various studies. (Charef et al., 2021) identified the lack of space as a barrier to implementing the circular economy in the construction industry. A case study Bangladesh (Ahmed & Sobuz, 2019) and Ghana (Amuna et al., 2021) found that the lack of space hindered the implementation of lean construction practices, emphasized the need for space to monitor manufacturing activities and quantify waste generation in industrialized construction facilities. These studies collectively demonstrate the difficulties and challenges associated with the lack of space in implementing consistent construction waste management practices.

3.3 Solutions

Construction and demolition waste management (CDWM) in the construction industry poses unique challenges requiring innovative solutions. The barriers mentioned above to effective

https://doi.org/10.52202/070182-0004

CDWM need comprehensive solutions that address the issues at various levels, from individual behaviors to government policies.
One possible explanation is increasing the stakeholder's knowledge and awareness about waste management practices and offering training on proper waste sorting and disposal techniques (Frempong-Jnr et al., 2023; Lu et al., 2015). Additionally, to enhance collaboration between stakeholders involved in CDWM, waste management committees or working groups can be established to facilitate coordination between building owners, waste management companies, and regulatory authorities, ensuring that waste management practices align with regulations and standards.(Kim et al., 2020). Through collaboration, stakeholders can develop more effective waste management strategies considering the unique challenges in the construction industry.

Secondly, the role of leadership and management is crucial in driving change and promoting sustainable waste management practices (Frempong-Jnr et al., 2023; Kim et al., 2020; Lu et al., 2015). Strong leadership commitment and top management support are essential for setting clear goals, allocating resources, and fostering a culture of waste reduction and recycling. Effective communication and engagement with employees and stakeholders can also create a sense of ownership and commitment to waste management initiatives.

Furthermore, the adoption of innovative technologies and practices, such as lean construction and supply chain management, can contribute to more efficient waste management (Ogunbiyi et al., 2014). Lean thinking focuses on waste reduction and process optimization, while supply chain management ensures effective coordination and collaboration among stakeholders involved in waste management. In addition, addressing technical hurdles and improving knowledge and expertise in waste management through training and education programs can help overcome implementation barriers (Kulatunga et al., 2006). This includes providing guidance on waste segregation, recycling techniques, and the use of modern technologies for waste tracking and management.

Waste management regulations and enforcement can be strengthened to ensure compliance with waste management practices. It can include the development of more stringent regulations for waste management and increasing enforcement of existing policies. Ensuring that all stakeholders comply with waste management regulations, the overall effectiveness of CDWM can be improved. Finally, allocating financial resources (Hassan et al., 2012; Hossain et al., 2017; Hwang & Bao Yeo, 2011; Karunasena & Amaratunga, 2016; Yuan et al., 2011) and providing incentives such as reduced waste disposal fees or tax credits (Osmani et al., 2008; Palafox-Alcantar et al., 2021; Yeheyis et al., 2013a) can motivate individuals and organizations to adopt proper waste management practices.

A good illustration of such regulations is introduced by the European Union, known as the "Waste Framework Directive (WFD) 2008/98/EC" (Spišáková et al., 2021; Zorpas et al., 2015). The directive sets forth principles for managing waste, notably the waste hierarchy that prioritizes waste prevention as the most preferred option, followed by Reuse, recycling, and recovery, with disposal being the least desirable choice (Spišáková et al., 2021)Therefore, member states must take action to manage construction and demolition (C&D) waste based on this hierarchy and establish national objectives for the recycling and recovery of C&D waste (Vieira & Pereira, 2015). Additionally, the regulation and policies established by the Environmental Protection Agency

https://doi.org/10.52202/070182-0004

(EPA) called (RCRA), and the "Circular Economy Promotion Law" governed by China are other notable examples of these activities (Aslam et al., 2020; Vieira & Pereira, 2015). The "Circular Economy Promotion Law" encourages resource efficiency and minimizes waste production. Construction enterprises must abide by prescribed regulations for waste management, which entail segregating and categorizing waste items, recycling reusable materials, and utilizing landfills only as a last option. While (RCRA) emphasizes setting standards for hazardous waste disposal (Aslam et al., 2020; Vieira & Pereira, 2015). These policies necessitate member states to initiate waste prevention measures, endorse the recycling and repurposing of construction waste, and guarantee proper handling and disposal of waste materials. Such standards aim to diminish the quantity of waste generated and enhance the efficient utilization of resources while minimizing the detrimental impact of debris on the environment.
It should be also mentioned that the effective strategy to mitigate the generation of Construction and Demolition (C&D) waste entails the implementation of regulations governing the design and construction of buildings and structures to facilitate disassembly and deconstruction (Rios et al., 2015). This approach enables companies to plan for the eventual removal of the building or construction at the end of its life cycle, thereby minimizing waste generated during demolition. Additionally, designing buildings for disassembly and deconstruction enables the salvage and reuse of materials, promoting a circular economy and reducing the demand for new materials. As such, this strategy provides a proactive solution to the challenges posed by the substantial environmental and economic costs associated with C&D waste management (Rios et al., 2015). By adopting these regulations, companies can enhance the sustainability of the construction industry, reduce waste generation, and promote a more environmentally conscious approach to infrastructure development.

In conclusion, overcoming implementation barriers in waste management strategies in the construction industry requires a comprehensive approach that involves collaboration, leadership, technological advancements, and regulatory support. By establishing waste management committees, promoting strong leadership, adopting innovative technologies, improving knowledge and expertise, and updating regulations, the construction industry in Canada can overcome barriers and achieve more effective and sustainable waste management practices.

4 Conclusion

The construction industry is a significant contributor to environmental concerns such as air and water pollution, depletion of natural resources, and increased greenhouse gas emissions. Sustainable waste management practices, such as reuse and recycling, present promising solutions to mitigate waste generation and promote sustainability.

The industry faces significant challenges in implementing sustainable practices effectively, including a lack of awareness, limited regulations and incentives, inadequate infrastructure, insufficient collaboration among stakeholders, and high expenses associated with waste management. To overcome these challenges, a multi-faceted approach is necessary, which includes education and awareness campaigns, policy and regulatory changes, investment in infrastructure and technology, and incentives for contractors and developers. Specialized waste

https://doi.org/10.52202/070182-0004

management infrastructure strengthened collaborations between stakeholders, and comprehensive evaluations of waste management systems are also necessary to promote sustainability.

It is crucial to take into account the social, economic, and environmental dimensions of sustainability when implementing sustainable waste management practices. Collaboration and communication between stakeholders are essential to promoting sustainability, and comprehensive evaluations of waste management systems are necessary to make informed decisions about system development. By adopting sustainable waste management practices, the construction industry can enhance sustainability, reduce waste generation, and promote a more environmentally conscious approach to infrastructure development. Future studies should explore the potential for innovative waste management practices such as the use of waste as a resource or the implementation of circular economy principles in the construction industry. There is a need for more research on the effectiveness of policies and regulations aimed at promoting sustainable waste management practices in the construction industry, as well as their impact on project costs and overall sustainability. Studies should also explore the potential for collaboration between different stakeholders in the construction industry to promote sustainable waste management practices and reduce waste generation.

References

Abarca-Guerrero, L., Lobo-Ugalde, S., Mendez-Carpio, N., Rodriguez-Leandro, R., & Rudin-Vega, V. (2022). Zero waste systems: Barriers and measures to recycling of construction and demolition waste. In *Sustainability* (Vol. 14, Issue 22). MDPI. https://doi.org/10.3390/su142215265

Ahmed, S., & Sobuz, M. H. R. (2019). Challenges of implementing lean construction in the construction industry in Bangladesh. *Smart and Sustainable Built Environment*, *9*(2), 174–207. https://doi.org/10.1108/SASBE-02-2019-0018

Ajayi, S. O., Oyedele, L. O., Akinade, O. O., Bilal, M., Owolabi, H. A., Alaka, H. A., & Kadiri, K. O. (2016). Reducing waste to landfill: A need for cultural change in the UK construction industry. *Journal of Building Engineering*, *5*, 185–193. https://doi.org/10.1016/j.jobe.2015.12.007

Ajayi, S. O., Oyedele, L. O., Bilal, M., Akinade, O. O., Alaka, H. A., & Owolabi, H. A. (2017). Critical management practices influencing on-site waste minimization in construction projects. *Waste Management*, *59*, 330–339. https://doi.org/10.1016/j.wasman.2016.10.040

Ajayi, S. O., Oyedele, L. O., Bilal, M., Akinade, O. O., Alaka, H. A., Owolabi, H. A., & Kadiri, K. O. (2015). Waste effectiveness of the construction industry: Understanding the impediments and requisites for improvements. *Resources, Conservation and Recycling*, *102*, 101–112. https://doi.org/10.1016/j.resconrec.2015.06.001

Asgari, A., Ghorbanian, T., Yousefi, N., Dadashzadeh, D., Khalili, F., Bagheri, A., Raei, M., & Mahvi, A. H. (2017). Quality and quantity of construction and demolition waste in Tehran. *Journal of Environmental Health Science and Engineering*, *15*(1), 14. https://doi.org/10.1186/s40201-017-0276-0

Aslam, M. S., Huang, B., & Cui, L. (2020). Review of construction and demolition waste management in China and USA. *Journal of Environmental Management*, *264*, 110445. https://doi.org/10.1016/j.jenvman.2020.110445

Bao, Z., Lee, W. M. W., & Lu, W. Implementing on-site construction waste recycling in Hong Kong: Barriers and facilitators. *Science of The Total Environment*, *747*, 141091. https://doi.org/10.1016/j.scitotenv.2020.141091

Bao, Z., & Lu, W. (2021). A decision-support framework for planning construction waste recycling: A case study of Shenzhen, China. *Journal of Cleaner Production*, *309*, 127449. https://doi.org/10.1016/j.jclepro.2021.127449

Frempong-Jnr, E. Y., Ametepey, S. O., & Cobbina, J. E. (2023). Impact of stakeholder management on efficient construction waste management. *Smart and Sustainable Built Environment*, *12*(3), 607–634. https://doi.org/10.1108/SASBE-08-2021-0147

Ghafourian, K., Kabirifar, K., Mahdiyar, A., Yazdani, M., Ismail, S., & Tam, V. W. Y. (2021). *Recycling*, *6*(4), 73. https://doi.org/10.3390/recycling6040073

Hao, J. L., Yu, S., Tang, X., & Wu, W. (2022). Determinants of workers' pro-environmental behaviour towards enhancing construction waste management: Contributing to China's circular economy. *Journal of Cleaner Production*, *369*. https://doi.org/10.1016/j.jclepro.2022.133265

Jin, R., Yuan, H., & Chen, Q. (2019). Science mapping approach to assisting the review of construction and demolition waste management research published between 2009 and 2018. *Resources, Conservation and Recycling*, *140*, 175–188. https://doi.org/10.1016/j.resconrec.2018.09.029

Kim, S. Y., Nguyen, M. V., & Luu, V. T. (2020). A performance evaluation framework for construction and demolition waste management: Stakeholder perspectives. *Engineering, Construction and Architectural Management*, *27*(10), 3189–3213. https://doi.org/10.1108/ECAM-12-2019-0683

Kulatunga, U., Amaratunga, D., Haigh, R., & Rameezdeen, R. (2006). Attitudes and perceptions of construction workforce on construction waste in Sri Lanka. *Management of Environmental Quality: An International Journal*, *17*(1), 57–72. https://doi.org/10.1108/14777830610639440

Llatas, C. (2011). A model for quantifying construction waste in projects according to the European waste list. *Waste Management*, *31*(6), 1261–1276. https://doi.org/10.1016/j.wasman.2011.01.023

Lu, W., Peng, Y., Webster, C., & Zuo, J. (2015). Stakeholders' willingness to pay for enhanced construction waste management: A Hong Kong study. *Renewable and Sustainable Energy Reviews*, *47*, 233–240. https://doi.org/10.1016/j.rser.2015.03.008

Manowong, E. (2012). Investigating factors influencing construction waste management efforts in developing countries: An experience from Thailand. *Waste Management & Research: The Journal for a Sustainable Circular Economy*, *30*(1), 56–71. https://doi.org/10.1177/0734242X10387012

Mendis, D., Hewage, K. N., & Wrzesniewski, J. (2015). Contractual obligations analysis for construction waste management in Canada. *Journal of Civil Engineering and Management*, *21*(7), Article 7. https://doi.org/10.3846/13923730.2014.893907

Mohammed, M., Shafiq, N., Elmansoury, A., Al-Mekhlafi, A.-B. A., Rached, E. F., Zawawi, N. A., Haruna, A., Rafindadi, A. D., & Ibrahim, M. B. (2021). https://doi.org/10.3390/su131910660

Mostafa, S., Chileshe, N., & Abdelhamid, T. (2016). Lean and agile integration within offsite construction using discrete event simulation: A systematic literature review. *Construction Innovation*, *16*(4), 483–525. https://doi.org/10.1108/CI-09-2014-0043

https://doi.org/10.52202/070182-0004

Ogunbiyi, O., Goulding, J. S., & Oladapo, A. (2014). An empirical study of the impact of lean construction techniques on sustainable construction in the UK. *Construction Innovation*, *14*(1), 88–107. https://doi.org/10.1108/CI-08-2012-0045

Ortiz, O., Castells, F., & Sonnemann, G. (2009). Sustainability in the construction industry: A review of recent developments based on LCA. *Construction and Building Materials*, *23*(1), 28–39. https://doi.org/10.1016/j.conbuildmat.2007.11.012

Osobajo, O. A., Oke, A., Omotayo, T., & Obi, L. I. (2022). A systematic review of circular economy research in the construction industry. *Smart and Sustainable Built Environment*, *11*(1), 39–64. https://doi.org/10.1108/SASBE-04-2020-0034

Palafox-Alcantar, P. G., Hunt, D. V. L., & Rogers, C. D. F. (2021).

Rios, F. C., Chong, W. K., & Grau, D. (2015). *118*, 1296–1304 https://doi.org/10.1016/j.proeng.2015.08.485

Saad, A., Bal, M., & Khatib, J. (2022). The Need for a Proper Waste Management Plan for the Construction Industry: A Case Study in Lebanon. *Sustainability*, *14*(19), 12783. https://doi.org/10.3390/su141912783

Shooshtarian, S., Caldera, S., Maqsood, T., Ryley, T., & Khalfan, M. (2022). An investigation into challenges and opportunities in the Australian construction and demolition waste management system. *Engineering, Construction and Architectural Management*, *29*(10), 4313–4330. https://doi.org/10.1108/ECAM-05-2021-0439

Shooshtarian, S., Maqsood, T., Wong, P. S. P., Khalfan, M., & Yang, R. J. (2020). Market development for construction and demolition waste stream in Australia. *Journal of Construction Engineering, Management & Innovation*, *3*(3), 220–231. https://doi.org/10.31462/jcemi.2020.03220231

Spišáková, M., Mésároš, P., & Mandičák, T. (2021). *Buildings*, *11*(2), 61. https://doi.org/10.3390/buildings11020061

Swetha, K. S., Tezeswi, T. P., & Kumar, M. V. N. S. (2022). Implementing construction waste management in India: An extended theory of planned behaviour approach. https://doi.org/10.1016/j.eti.2022.102401

Tam, V. W. Y. (2009). Comparing the implementation of concrete recycling in the Australian and Japanese construction industries. *Journal of Cleaner Production*, *17*(7), 688–702. https://doi.org/10.1016/j.jclepro.2008.11.015

Tam, V. W. Y., Hai, S., & Tam, C. M. (2005). Reuse, recycle and reduction of construction and demolition waste (C&D Waste): A Hong Kong study. *Proceedings of Criocm 2005 International Research Symposium on Advancement of Construction Management and Real* –655 https://www.webofscience.com/wos/woscc/full-record/WOS:000235423100088

Tsai, F. M., Bui, T.-D., Tseng, M.-L., & Wu, K.-J. (2020). A causal municipal solid waste management model for sustainable cities in Vietnam under uncertainty: A comparison. *Resources, Conservation and Recycling*, *154*, 104599. https://doi.org/10.1016/j.resconrec.2019.104599

Vieira, C. S., & Pereira, P. M. (2015). Damage induced by recycled Construction and Demolition Wastes on the short-term tensile behaviour of two geosynthetics. *Transportation Geotechnics*, *4*, 64–75. https://doi.org/10.1016/j.trgeo.2015.07.002

Wu, Z., Yu, A. T. W., & Poon, C. S. (2020). Promoting effective construction and demolition waste management towards sustainable development: A case study of Hong Kong. *Sustainable Development*, *28*(6), 1713–1724. https://doi.org/10.1002/sd.2119

Yeheyis, M., Hewage, K., Alam, M. S., Eskicioglu, C., & Sadiq, R. (2013a). An overview of construction and demolition waste management in Canada: A lifecycle analysis approach to sustainability. *Clean Technologies and Environmental Policy*, *15*(1), 81–91. https://doi.org/10.1007/s10098-012-0481-6

Yeheyis, M., Hewage, K., Alam, M. S., Eskicioglu, C., & Sadiq, R. (2013b). An overview of construction and demolition waste management in Canada: A lifecycle analysis approach to sustainability. *Clean Technologies and Environmental Policy*, *15*(1), 81–91. https://doi.org/10.1007/s10098-012-0481-6

Yu, A. T. W., Wong, I., & Mok, K. S. H. (2021). *Environmental Challenges*, *5*, 100231. https://doi.org/10.1016/j.envc.2021.100231

Zorpas, A. A., Lasaridi, K., Voukkali, I., Loizia, P., & Chroni, C. (2015). Promoting Sustainable Waste Prevention Strategy Activities and Planning in Relation to the Waste Framework Directive in Insular Communities. *Environmental Processes*, *2*(S1), 159–173. https://doi.org/10.1007/s40710-015-0093-3

https://doi.org/10.52202/070182-0004

Navigating the future of Project Management: Exploring emerging technologies and the dynamic landscape of evolving leadership in Construction

Gurneet Singh Sidhu[1*], Adel A. Zadeh[2]
[1*]College of Professional Studies, Northeastern University, Toronto, Canada; Sidhu.g@northeastern.edu
[2] College of Professional Studies, Northeastern University, Toronto, Canada; a.zadeh@northeastern.edu
* Correspondence:Sidhu.g@northeastern.edu

Abstract

The world of project management is witnessing the beginning of a revolution due to emerging technologies, the most relevant being Artificial Intelligence (AI), Machine Learning (ML), Natural Language Processing (NLP), and Big Data. This research paper examines the impact of these technologies in the project management domain, with a focus on the construction industry. With the increasing complexity and scale of construction projects, there is a growing need to leverage technology to optimize project execution, resource allocation, and decision-making processes. The research methodology employed in this study incorporates an in-depth literature review and a survey to examine the challenges and benefits associated with the implementation of these technologies in the construction industry. The findings suggest that AI-driven technologies such as ML, NLP, and predictive analytics offer opportunities to enhance planning and scheduling, optimize resource allocation, automate repetitive tasks, mitigate risks, and improve project performance. This study also highlights the importance of the Project Management Technology Quotient (PMTQ) as a measure of an individual's or organization's ability to adapt and leverage technology effectively, as well as the role ambidextrous leadership can play in shaping the new era of technology.

Keywords: Project Management Technology Quotient (PMTQ), ambidextrous leadership, artificial intelligence, construction management

1 Introduction

The Project Management Technology Quotient (PMTQ) has gained significant interest in the Project Management Community, defined as a measure of an organization's ability to leverage innovation and adopt disruptive technologies for successful project outcomes (Zaman et al., 2020). The Project Management Institute (PMI) reports that PMTQ is gaining more and more importance as individuals and businesses pursue digital sustainability (PMI, 2020; Zaman et al., 2020). According to Gartner's research, by 2030, 80% of project management tasks will be conducted by AI, powered by big data, ML, and NLP (Gartner, 2019; Nieto-Rodriguez & Vargas, 2023). Researchers and entrepreneurs have already developed algorithms to apply AI and ML to the world of project management, and when the next generation of project management tools is widely adopted, profound changes are bound to occur. These technological innovations will disrupt six aspects of project management, including improved selection and prioritization, support for the project management office (PMO), quicker project definition, planning, and reporting, and virtual project assistants (Nieto-Rodriguez & Vargas, 2023). These tools will aid leaders and project

https://doi.org/10.52202/070182-0005

managers in anticipating risks, facilitating project definition, planning, and reporting, and delivering real-time information on the project's status, benefits, slippage, and team sentiment (Nieto-Rodriguez, 2021). The PMI 2020 Pulse of the Profession® revealed that enterprise-wide adoption of emerging technologies is making project managers more critical leaders (PMI, 2020). Being tech savvy will be crucial, and project managers must enhance their technical methods of operation. AI will not administer the project on its own, but it will free up valuable resources to focus on work with greater value (Chakravorti, 2022). The allocation of time will shift as AI enables more value-driven work and machine learning tools replace programmable functions. As AI and other advanced technologies grow in importance, project managers will need to acquire proficiency in areas where technology falls short. The PMI identifies collaboration and problem-solving as indispensable talents for project leaders (PMI, 2020).

For the successful application of emerging technologies to projects and project management practices, factors such as accuracy, investment, willingness to abandon old habits, training, adaptability, and the executive sponsor's and senior leaders' abilities are required (Brassey et al., 2021). By eliminating procedural inefficiencies, avoiding surprises, and effectively mitigating risks, transformational leadership and technology integration can significantly contribute to innovation success. Innovators who integrate emerging technologies into their respective disciplines are receptive, possess a healthy degree of cynicism, and are aware of when to assimilate the emerging technologies.

An inclusive leadership approach manages people and technology effectively, cultivating a culture in which people efficiently manage technology to generate greater business value. Innovation success is a framework for measuring organizational performance and innovation strategy outcomes. Transformational leaders have always exhibited indispensable managerial qualities, fostering revolutionary change and technological advancements across industries by establishing new and inspiring directions, resulting in substantial innovation success (Zaman et al., 2020). Hence, the future of project management is swiftly transforming, with AI and ML technologies reshaping numerous aspects of project management. These tools will significantly reduce the amount of time spent on tiresome testing tasks, decrease the number of reworks, and produce solutions that are both user-friendly and bug-free.

The application of AI to project management will yield significant benefits, including the automation of administrative and low-value tasks as well as enhanced project selection, definition, and execution. It is believed that the construction industry is sluggish to adopt technological innovations and eliminate obstacles from the past. This has resulted in exceedingly low levels of development and productivity. Even the most organized project managers can become overwhelmed by the plethora of stakeholders, extensive documentation, labor management, change orders, and modifications involved. Emerging technologies, such as AI, have a great deal of potential to improve the construction industry (Rathod & Sonawane, 2022; Regona et al., 2022). A few studies have investigated the advantages of AI and other technologies in the construction industry. As such AI is reported to have the potential to significantly enhance the efficacy of construction project management and play a significant role in knowledge synthesis to facilitate informed decision-making in the construction industry (Srivastava et al., 2022).

 https://doi.org/10.52202/070182-0005

Due to its limited practical application, the current study is exploratory in nature. This paper also undertakes a literature review of emerging trends in technology-inclusive project management and discusses the implications of integrating AI technology into construction projects.

2 Literature on emerging technologies in the construction industry

This section aims to review the literature on the impact of emerging technologies on project management and leadership. It analyzes relevant research publications and highlights key characteristics and trends, as well as their potential implications for project management processes and leadership practices. This section will help to comprehend the current landscape of emerging technologies in project management and leadership by synthesizing the existing literature.
Organizations are experiencing an explosion in data, both structured and unstructured, which has led to the development of big data analytics (BDA) in various applications (Adnan et al., 2021). The increasing use of big data in various industries has led to analytic capabilities and processes for redefining innovation, competition, and productivity across industries (Ahuja, 2021). The application of analytic processes and skills is required to convert massive amounts of data into valuable insights, thereby enhancing project and organizational decision-making.

A study by Ahmed et al. (2022) validates the fundamental functions of human capabilities (decision-making) and technological capabilities (big data analytics) in enhancing competitive advantage by implementing projects in a big data environment. The relationship between big data analytics and successful project completion was analyzed. The research discovered a correlation between project achievement, big data analytics, and decision making.

Social research methods and data mining techniques can help understand the impact of big data on project management. Understanding big data privacy and the role of big data in project management can help organizations navigate the challenges of managing large data sets and achieve project maturity (Williams et al., 2014). The growing complexity of manufacturing companies' processes and their impact on decision-making make it essential to integrate the evolving technologies of the Internet of Things (IoT) into project management applications (Atzori et al., 2018). The IoT is defined as a dynamic global network infrastructure with self-configuring capabilities based on standard and interoperable communication protocols (Murray et al., 2016)(ref). IoT components can sense and react to real-world events, providing accurate and current information services almost in real-time. This information is useful in complex projects that need to perform timely reactions to selected events (Ghimire et al., 2016). AI technologies are expected to significantly impact project management on multiple levels. Holzmann et al. (2022) addressed this topic through a Delphi study with 52 project management experts who reflected on future potential AI applications for the project management knowledge areas. The findings revealed that AI applications are expected to improve decision-making processes, save time and money, and enhance interpersonal and strategic activities. Project managers recognize the need to resolve fundamental processes, but they are also receptive to advanced technologies that enhance their leadership abilities. Applications of AI will complement human activities in decision-making situations, particularly in dubious, complex, and ambiguous environments.

Srivastava et al.(2022) demonstrated the significance of automation in the construction sector and found that technologies such as AI, IoT, big data, 4D printing, and cloud computing will increase the opportunities for widespread implementation of automation in construction in order to create

https://doi.org/10.52202/070182-0005

sustainable and resilient infrastructure. In another study, Ramalingham & Mossman (2022) conducted qualitative and quantitative research and concluded that it is unlikely that AI will replace human labor. However, it will improve project delivery by reducing costly errors, enhancing safety precautions, and enhancing the efficiency of building operations. Apart from that, through the BIM platform, new employment rolls are also coming into the business, such as BIM managers, leaders, and coordinators.

3 Methodology

The study began with an extensive literature review, which aimed to analyze the influence of emerging technologies on project management and the subsequent evolution of leadership in this context. This review encompassed a wide range of literature sources to establish a comprehensive understanding of the subject. As the research progressed, the focus shifted towards exploring emerging technologies with the highest potential for future impact. In particular, the study delved into one of the most intricate domains of project management, namely construction.

By narrowing down the scope, the authors conducted an in-depth examination of how emerging technologies could transform and enhance project management practices within the construction industry. Through this approach, the study aimed to provide valuable insights into the specific challenges and opportunities associated with the adoption of emerging technologies in construction project management. By scrutinizing this complex field, the authors sought to identify the potential benefits, limitations, and strategies for effectively integrating and leveraging emerging technologies to improve project outcomes in the construction sector. Given the novelty of the topic within the construction industry, this study employed an exploratory approach. The combination of cutting-edge technology, specifically AI, with traditional project delivery techniques was explored through an intuitive and inductive questioning style.

In this study, a survey was conducted using Google Forms to examine the complexities within the construction sector and explore the potential solutions that AI can offer. The survey gathered responses from a total of 40 participants, including professionals and students affiliated with the construction industry. The participant pool encompassed employees from various construction firms and graduate-level students from esteemed institutions such as the National Institute of Technology, Hamirpur, Northeastern University in Toronto, RMIT in Australia, UNSW Sydney, and Pratt Institute in Brooklyn, New York.

To construct the survey questionnaire, the authors conducted interviews with civil engineers, architects, and project managers from SJVN Ltd. and Patel Infrastructure. These industry experts provided valuable insights based on their experience with construction projects. The research questions were formulated with the intention of addressing the industry's challenges and exploring the potential solutions that AI can currently provide or may offer in the near future. The findings revealed that the primary challenges faced by the construction industry, as identified by both firms, were rooted in traditional practices and a reluctance to embrace automation, AI, and advanced construction management techniques. Subsequently, the survey questionnaire was distributed to the participating firms and students to gather a broader range of opinions on the topic.

https://doi.org/10.52202/070182-0005

4 Results

As shown in Figure 1, the survey data revealed several key reasons for the shortage of labor in the construction industry. Among the 40 participants, 47.5% attributed the shortage to low wages, indicating that the current compensation levels are not competitive enough to attract and retain skilled workers. Additionally, 25% of respondents cited a high number of accidents as a significant factor contributing to the labor shortage, highlighting that safety concerns and a perception of the construction industry as a hazardous working environment may deter potential workers from entering the field. Furthermore, 45% of participants pointed to extreme working conditions as another critical factor impacting the availability of labor. These conditions may encompass various challenges, such as long working hours, exposure to adverse weather conditions, and physically demanding tasks. Finally, the data showed that 57.5% of respondents believed that the ratio of raw manual work to reward was imbalanced, implying that the effort required in construction jobs may not be adequately compensated, discouraging potential workers from pursuing careers in the industry.

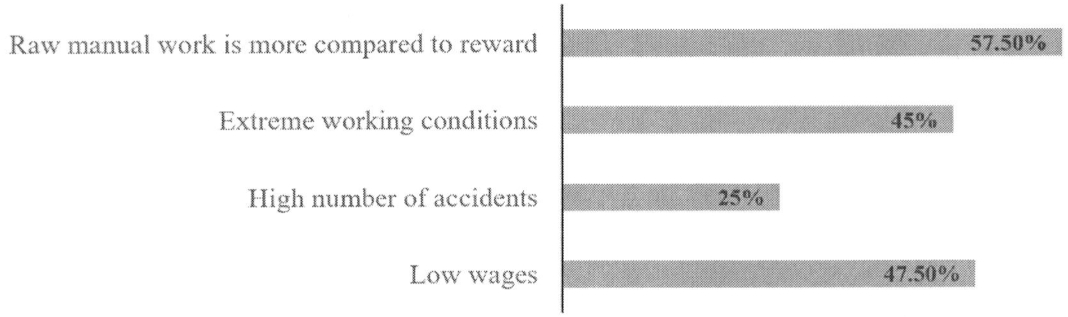

Figure 1. Factors contributing to the labor shortage in the construction industry

With regards to major problems faced in the construction industry (Figure 2), the most cited issue is project delay, with a significant 85% of respondents highlighting it as a major problem. Delays in construction projects can have far-reaching consequences, leading to increased costs, dissatisfied clients, and potential legal disputes (Ibrahim et al., 2022; Mohammed et al., 2021). Cost overrun is another prominent problem, with 62.5% of participants acknowledging its significance, and it can have severe financial implications for both contractors and clients. Proper cost estimation, resource management, and risk assessment are crucial aspects that need to be addressed to mitigate cost overrun challenges in the construction industry (Aljohani, 2017; Skorupka, 2008).

30% of respondents mentioned disputes as a significant issue in the construction industry. Disputes can arise due to various reasons, including contractual disagreements, design issues, delays, or quality disputes (Alaloul et al., 2019; El-Sayegh et al., 2020). Safety concerns for workers were reported by 37.5% of participants. This is not unexpected since the construction industry is known for its inherent risks, and ensuring the safety and well-being of workers should be a paramount concern for all stakeholders (Al-Bayati, 2021). The integration of emerging technologies and innovative management strategies may provide opportunities to tackle these challenges and drive positive changes in the construction industry.

https://doi.org/10.52202/070182-0005

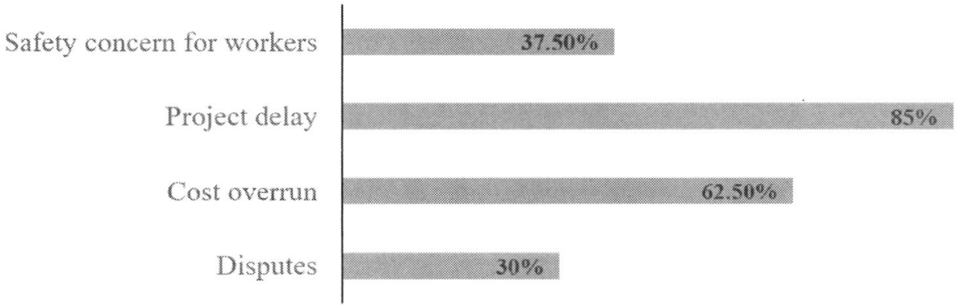

Figure 2. Major challenges encountered in the construction industry

The survey data also sheds light on the reasons why construction projects exceed their budget and planned timeline (Figure 3). When it comes to exceeding the budget (Figure 3a), the data indicates that scope creep is the most cited factor, with 22.5% of respondents attributing budget overruns to the uncontrolled expansion of project scope beyond its original boundaries, leading to additional costs. Poor estimation techniques are another important factor, according to 27% of participants, which suggests that inaccurate cost estimates during the project planning phase can lead to financial discrepancies during project execution. Additionally, 16% of respondents highlighted the lack of equipment on-site as a contributing factor to exceeding the budget. The absence or delay of essential construction equipment can lead to project delays and increased costs. Lastly, 35% of participants pointed to poor surveying techniques as a reason for budget overruns. This implies that inaccurate site assessments and measurements can lead to unexpected expenses and resource inefficiencies.

Regarding construction project delays (Figure 3b), the primary reason identified in the survey is the lack of proper construction management techniques, which was reported by a significant 80% of respondents. This suggests that deficiencies in project planning, scheduling, and execution are major factors contributing to delays. Improper resource levelling was also seen as a significant factor, with 47.5% of participants highlighting it as a cause of project schedule lags. Resource levelling involves optimizing resource utilization to prevent overallocation and under allocation, which can impact project timelines. Labor shortage was cited by 32.5% of respondents, indicating that a lack of skilled workers in the construction industry can hinder project progress. Additionally, 22.5% of participants mentioned improper contract formation as a reason for schedule delays, implying that unclear or poorly structured contracts can lead to disputes and project disruptions.

https://doi.org/10.52202/070182-0005

(a)

(b)

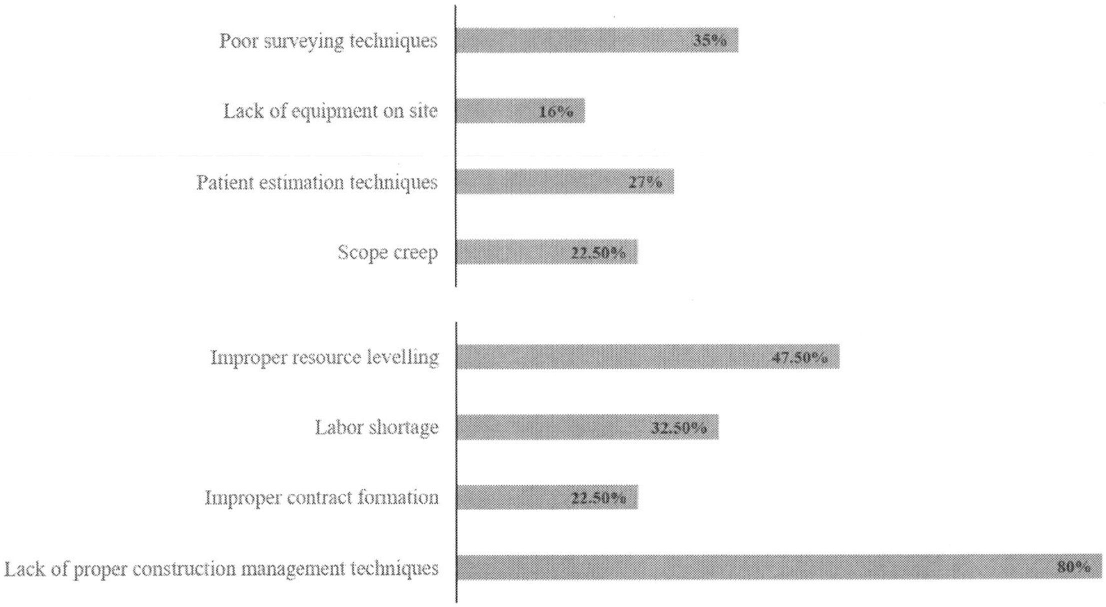

Figure 3. Factors contributing to construction projects (a) exceeding budget (b) lagging behind schedule

To fast-track construction projects and meet schedule deadlines, survey respondents provided valuable insights into effective strategies (Figure 4a). Firstly, 70% of participants recommended adopting AI-induced project management methodologies. This suggests that leveraging artificial intelligence in project planning, scheduling, and resource allocation can significantly enhance project efficiency and streamline processes, leading to quicker project completion. Similarly, 67.5% of respondents highlighted the importance of automation in construction. Automation can reduce manual labor, improve productivity, and expedite various construction tasks, thus contributing to faster project delivery (Son et al., 2010). Additionally, 27.5% of participants suggested using the Internet of Things (IoT) to fast-track construction projects. IoT technology can enable real-time monitoring of construction activities, optimize equipment usage, and facilitate better coordination, all of which can lead to faster and more efficient project execution (W. Pan et al., 2017). Finally, an equal number of respondents, 27.5%, recommended increasing labor at the construction site.

To resolve the problem of structures collapsing globally, survey respondents provided the following insights (Figure 4b): An overwhelming 80% of participants advocated for strict laws for quality control. Implementing and enforcing stringent regulations can ensure that construction projects adhere to high-quality standards, thereby reducing the risk of structural failures. Moreover, 57.5% of respondents suggested adopting artificial intelligence-aided design software. Furthermore, 17.5% of participants recommended finding better civil engineers. Skilled and experienced engineers can provide expertise and meticulous planning to prevent structural failures. Only 5% of respondents mentioned using alternative construction materials as a resolution. While innovative materials can enhance construction safety, it seems that the majority of participants focused on other strategies for addressing the issue.

https://doi.org/10.52202/070182-0005

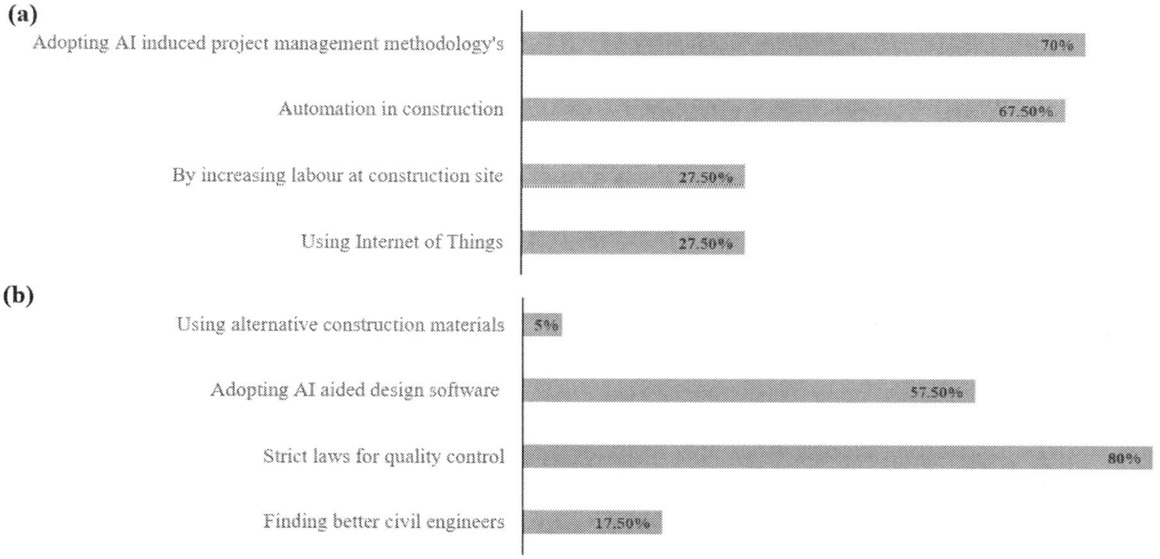

Figure 4. (a) Strategies for fast-tracking construction projects to meet schedule deadlines (b) addressing global structural collapse solutions

Based on the survey responses, it is evident that organizations are extensively leveraging artificial intelligence in various aspects of their operations (Figure 5a).

For planning and designing, a significant 80% of respondents reported using generative design through AI. Generative design allows the organization to explore numerous design options quickly, optimizing for various parameters and finding innovative solutions (Qi et al., 2021).

In fleet management, 27.5% of participants indicated utilizing AI. AI in fleet management can optimize vehicle routes, monitor vehicle health, and enhance overall efficiency in transportation.

For measuring site progress, 45% of respondents reported making use of AI. AI technologies can facilitate automated data collection, image recognition, and analysis to track construction progress accurately. Similarly, 32.5% of participants mentioned employing AI for safety management. AI can be utilized to identify potential hazards, monitor safety compliance, and enhance overall safety protocols on construction sites. Regarding alternative labor shortages, 15% of respondents reported using AI-based solutions. AI can help automate certain tasks and processes, filling in gaps caused by labor shortages and improving productivity.

When it comes to AI-based software (Figure 5b), Building Information Modelling (BIM) was the most popular, with 65% of participants using it. BIM facilitates comprehensive project visualization, collaboration, and coordination, enhancing project efficiency. Geospatial AI and Revit both received 30% of the responses. Geospatial AI allows for advanced spatial analysis and decision-making based on geographical data, while Revit is a popular BIM software for designing and constructing buildings. Procore and Smartsheet were used by 7.5% and 30% of respondents, respectively. Procore is a construction management platform that leverages AI for project insights, while Smartsheet is a project management tool that can benefit from AI-powered automation.

The majority of respondents (92.5%) strongly believe that AI has the potential to revolutionize the construction industry in the future (Figure 5c). This overwhelming response suggests a high level of confidence in AI's capacity to bring transformative changes and innovation to the construction sector.

https://doi.org/10.52202/070182-0005

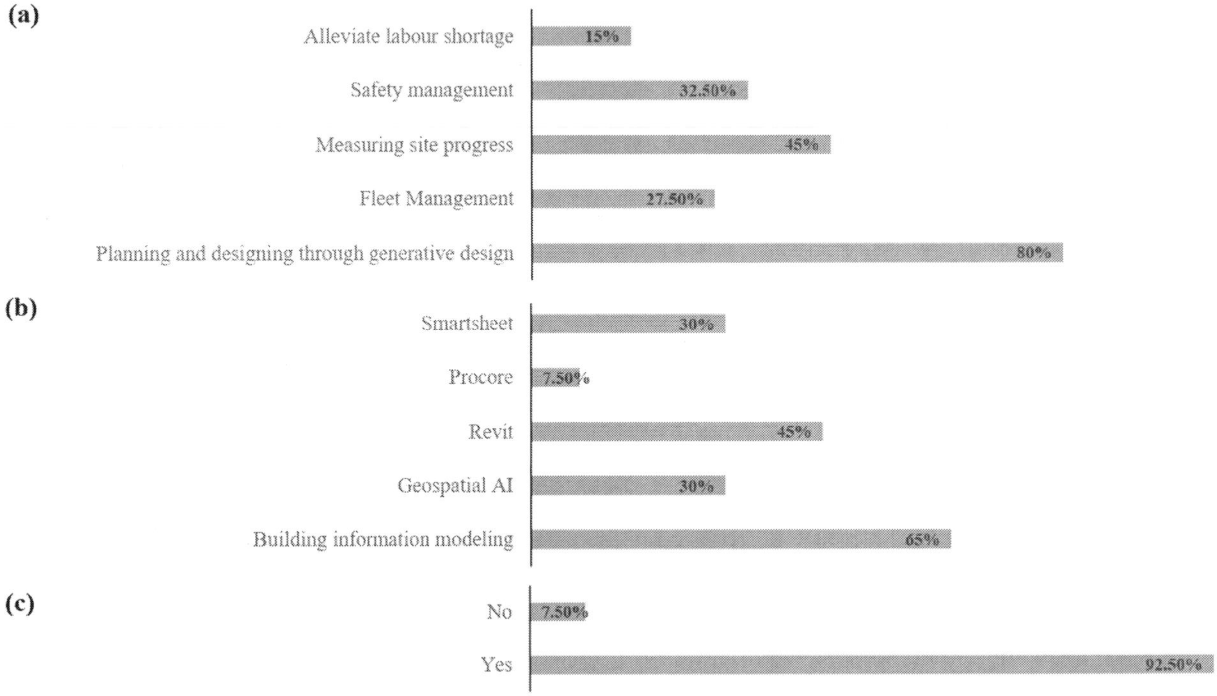

Figure 4. Insights into (a) AI Implementation in the organization, (b) utilized AI-based software, and (c) the potential of AI in revolutionizing the construction industry

5 Discussion

AI plays a crucial role in the entire life cycle of a project, from planning to execution, operations, and maintenance phases (Arayici et al., 2011; Zaman et al., 2020). Traditional planning techniques, such as the Critical Path Method (CPM), are ineffective for delivering projects on time and within budget due to reactive control measures and hand-off approaches (Boyacioglu et al., 2022). Advanced collaborative tools such as Building Information Modelling (BIM) are designed to reduce revisions by simulating the construction sequence in advance and detecting trade conflicts (Arayici et al., 2011). AI can reduce the time needed to develop the required skills and integrate with BIM to enable generative design solutions (Ma et al., 2021). Nevertheless, some contend that CPM cannot facilitate end-to-end synchronization and coordination of transactions (Boyacioglu et al., 2022).

In the construction industry, trade-specific technical knowledge and local knowledge are primarily implicit. Organization theorists contend that AI-powered, centralized, and controlled planning might outperform decentralized interactions in coordinating economic activity. AI and advanced digital tools are notable for maintaining site logistics, enhancing productivity, quality, and safety, and assessing project performance and risks (Bang & Nils, 2022). The Last Planner System (LPS) of the Lean Construction Institute focuses on crew leaders managing production collaboratively with project superintendents, enabling production flow and early identification and removal of

https://doi.org/10.52202/070182-0005

production constraints and bottlenecks (Ramalingham & Mossman, 2022) AI and machine learning (ML) can help assess and manage the operation and performance of buildings, such as intelligent buildings (IBs), which help building owners, property managers, and occupants realize their goals in areas of cost, comfort, convenience, safety, long-term flexibility, and marketability.

AI can significantly improve construction project management effectiveness, with potential applications in the construction industry forecasted to reach $4.51 billion by 2026 AI can help predict cost variance based on factors such as project size, contract type, and project manager experience level, and help staff get training material remotely (Souhail et al., 2020). AI can also monitor possible risks on job sites, allowing contractors to come up with more effective contingency plans. Construction firms are investing in AI and data science to improve productivity and address labor shortages. Real-time analysis of data suggests that companies could increase productivity by up to 50%. AI can also help robots perform repetitive tasks more efficiently, freeing up human workers for other construction activities. AI has the potential to improve quality, enhance safety, and reduce costs within the construction workflow. AI and digital transformation are transforming the construction industry by providing convenience and safety for workers. Job safety analysis and safety programs have improved since the early 1900s, with AI playing a significant role in reducing construction costs (Souhail et al., 2020). Algorithms for machine learning have aided the decision-making process by facilitating autonomous data analysis and decision-making. Drones and sensors facilitate human less construction monitoring on-site by providing a comprehensive picture of the site throughout the various project phases. Monitoring, recognizing, evaluating, and predicting potential risks in terms of safety, quality, efficiency, and cost across teams and work areas is possible with AI-based risk mitigation. Process mining is another important use of AI techniques in optimization problems. Digitalization has played a significant role in the construction industry, with BIM playing a leading role in digitalizing the construction process. IoT, a new generation of IoT, incorporates AI techniques into IoT infrastructure for more efficient operation and data analysis. The synergy between IoT and BIM under 5G wireless communication is expected to significantly promote data collection, transmission, and processing in the smart home, smart city, and smart construction industries.4D printing is an emerging technology that adds the fourth dimension of time to 3D printing, focusing on intelligent behavior and self-assembly. Blockchain technology can improve project management systems in construction, smart energy, cities, government, homes, and transportation. It can improve built asset sustainability, automate project delivery, and reduce payment fraud by creating smart contracts Smart robotics is rapidly advancing in construction applications, with ground robots automating repetitive tasks and aerial robots using unmanned aerial vehicles (UAVs) for land survey, site monitoring, and structure health monitoring. VR/AR has been applied in hazardous construction scenarios, engineering education, and training to enhance cognitive learning and safety consciousness (Y. Pan & Zhang, 2021).

5.1 Ideal Leadership for Construction Industry

Ambidextrous leadership promotes innovative work behavior, as it encourages individuals to upgrade their knowledge and enhance work performance both at the individual and organizational levels. Ambidextrous leadership is a crucial concept in leadership theory, as it allows leaders to balance transformational and transactional behavior in different situations. There are three characteristics of ambidextrous leadership: opening leadership practices, closing leadership practices, and a malleable characteristic that can select the best one based on the situation. These

https://doi.org/10.52202/070182-0005

practices can foster all four aspects of innovative work behavior (IWB): opportunity exploration, idea cultivation, idea promotion, and idea recognition. Innovation requires experimentation, thinking outside the box, and imagination. Leaders play an essential role in the generation of new ideas by supporting and encouraging them and convincing them of the importance of novel ideas for the innovation process. Ambidextrous leadership has a positive impact on knowledge-sharing, which is an essential element of organizational success. Knowledge management (KM) involves securing, sharing, and applying knowledge, which is crucial for organizational learning and innovation. By promoting IWB and gaining knowledge from other employees' experiences, organizations can encourage the process of implementing emerging technologies and improve productivity in construction (Haider et al., 2023).

6 Conclusion

The integration of AI and other emerging technologies in project management, including the construction industry, holds tremendous promise for revolutionizing project execution, optimizing resource allocation, and delivering successful outcomes. AI is not yet a standard tool for project management, but the technology will undoubtedly disrupt the field in the near future. AI and machine learning can significantly outperform humans at making predictions and identifying patterns that cannot be discerned by any other means. The literature review indicates that AI, along with other emerging technologies such as big data analytics, machine learning, and natural language processing, can automate administrative tasks, enhance decision-making processes, and improve overall project outcomes. However, it is important to note that while there is considerable potential, the construction industry, in particular, may face unique challenges in adopting and integrating these technologies due to factors like multiple stakeholders, extensive documentation, and labor management. By addressing these challenges and leveraging the opportunities presented by AI integration, the construction industry can streamline operations, reduce costs, and improve project efficiency.

Emerging technologies have created a situation of dilemma for project managers, as they fear that automation of a significant part of their tasks might create a scarcity of professional opportunities in the near future. The best project managers will learn to use these tools to their advantage, shifting their focus from tedious and repetitive technical aspects to planning more value-added activities, thereby increasing the success rate of their projects. Future project managers will require exceptional interpersonal skills, leadership abilities, strategic thinking, and business savvy. The Project Management Technology Quotient (PMTQ) has emerged as a measure of an individual's or organization's ability to adapt and leverage technology effectively in project management practices. As technology continues to advance rapidly, project managers and organizations should invest in developing their PMTQs and embrace the integration of emerging technologies to stay competitive. This will require individuals to be flexible and adaptable to the sector's cultural shift. In essence, the sustenance of AI would depend more on its complementarity with human skills than on its capacity to replace them. Both qualitative and quantitative validation studies are required to support the empirical findings. Continued research, experimentation, and application in the real world will be required to completely exploit the benefits and overcome the obstacles associated with AI integration in project management.

https://doi.org/10.52202/070182-0005

References

Adnan, K., Akbar, R., & Wang, K. S. (2021). Development of usability enhancement model for unstructured big data using SLR. *IEEE Access*, *9*, 87391–87409. https://doi.org/10.1109/ACCESS.2021.3089100

Ahuja, T. (2021). Leveraging the big data tools and techniques to attempt scientific analytics for drawing an optimum business intelligence. *International Journal of Research in Science and Technology*, *11*(04), 20–25. https://doi.org/10.37648/ijrst.v11i04.003

Alaloul, W. S., Hasaniyah, M. W., & Tayeh, B. A. (2019). A comprehensive review of disputes prevention and resolution in construction projects. *MATEC Web of Conferences*, *270*, 05012. https://doi.org/10.1051/matecconf/201927005012

Al-Bayati, A. J. (2021). Impact of construction safety culture and construction safety climate on safety behavior and safety motivation. *Safety*, *7*(2), 41. https://doi.org/10.3390/safety7020041

Aljohani, A. (2017). Construction projects cost overrun: what does the literature tell us? *International Journal of Innovation, Management and Technology*, 137–143. https://doi.org/10.18178/ijimt.2017.8.2.717

Arayici, Y., Coates, P., Koskela, L., Kagioglou, M., Usher, C., & O'Reilly, K. (2011). Technology adoption in the BIM implementation for lean architectural practice. *Automation in Construction*, *20*(2), 189–195. https://doi.org/10.1016/j.autcon.2010.09.016

Atzori, L., Lera, A., & Morabito, G. (2018). The internet of things : a survey. *Tạp Chí Nghiên Cứu Dân Tộc*, *24*. https://doi.org/10.25073/0866-773X/64

 https://doi.org/10.52202/070182-0005

Bang, S., & Nils, O. (2022). Artificial intelligence in construction projects: a systematic scoping review. *Journal of Engineering, Project, and Production Management.* https://doi.org/10.32738/JEPPM-2022-0021

Boyacioglu, S. E., Greenwood, D., & Rogage, K. (2022). Incorporating emerging technologies in the forensic analysis of construction project delays. *IOP Conference Series: Earth and Environmental Science, 1101*(5), 052029. https://doi.org/10.1088/1755-1315/1101/5/052029

Brassey, J., De Smet, A., Kothari, A., Lavoie, J., Mugayar-Baldocchi, M., & Zolley, S. (2021). Future proof: solving the 'adaptability paradox'for the long term. *McKinsey: Our Insights*.

Chakravorti, B. (2022, March 17). Why AI failed to live up to Its potential during the pandemic. *Harvard Business Review*. https://hbr.org/2022/03/why-ai-failed-to-live-up-to-its-potential-during-the-pandemic

El-Sayegh, S., Ahmad, I., Aljanabi, M., Herzallah, R., Metry, S., & El-Ashwal, O. (2020). Construction disputes in the UAE: causes and resolution methods. *Buildings*, *10*(10), 171. https://doi.org/10.3390/buildings10100171

Gartner. (2019). Project management tasks to be eliminated as AI takes over | Gartner. *https://www.gartner.com/en/newsroom/press-releases/2019-03-20-gartner-says-80-percent-of-today-s-project-management*

Ghimire, S., Luis-Ferreira, F., Nodehi, T., & Jardim-Goncalves, R. (2016). IoT based situational awareness framework for real-time project management. *International Journal of Computer Integrated Manufacturing*, 1–10. https://doi.org/10.1080/0951192X.2015.1130242

Haider, S. A., Zubair, M., Tehseen, S., Iqbal, S., & Sohail, M. (2023). How does ambidextrous leadership promote innovation in project-based construction companies? Through mediating role of knowledge-sharing and moderating role of innovativeness. *European Journal of Innovation Management*, *26*(1), 99–118. https://doi.org/10.1108/EJIM-02-2021-0083

Holzmann, V., Zitter, D., & Peshkess, S. (2022). The expectations of project managers from artificial intelligence: a delphi study. *Project Management Journal*, *53*(5), 438–455. https://doi.org/10.1177/87569728211061779

Ibrahim, A. T., Sirajo, A., & Ibrahim, A. H. (2022). The effects of delay and time management building construction: a review. *International Journal Of Scientific Advances*, *3*(4). https://doi.org/10.51542/ijscia.v3i4.29

Ma, W., Wang, X., Wang, J., Xiang, X., & Sun, J. (2021). Generative design in building information modelling (BIM): approaches and requirements. *Sensors*, *21*(16), 5439. https://doi.org/10.3390/s21165439

Mohammed, A., Almousa, A., Ghaithan, A., & Hadidi, L. A. (2021). The role of blockchain in improving the processes and workflows in construction projects. *Applied Sciences*, *11*(19), 8835. https://doi.org/10.3390/app11198835

Murray, A., Papa, A., Cuozzo, B., & Russo, G. (2016). Evaluating the innovation of the internet of things: empirical evidence from the intellectual capital assessment. *Business Process Management Journal*, *22*(2), 341–356. https://doi.org/10.1108/BPMJ-05-2015-0077

Nieto-Rodriguez, A. (2021). Harvard business review project management handbook: How to launch, lead, and sponsor successful projects. *Harvard Business Review Press.*

Nieto-Rodriguez, A., & Vargas, R. V. (2023, February 2). How AI will transform project management. *Harvard Business Review*. https://hbr.org/2023/02/how-ai-will-transform-project-management

Pan, W., Langosch, K., & Bock, T. (2017, July 1). Development of the process information modelling in the construction project: a case study of the zero-plus project. *34th International Symposium on Automation and Robotics in Construction, Taipei, Taiwan.* https://doi.org/10.22260/ISARC2017/0108

Pan, Y., & Zhang, L. (2021). Roles of artificial intelligence in construction engineering and management: a critical review and future trends. *Automation in Construction*, *122*, 103517.

PMI. (2020). Ahead of the curve: forging a future-focused culture. (*Pulse of the Profession*).

Qi, B., Razkenari, M., Costin, A., Kibert, C., & Fu, M. (2021). A systematic review of emerging technologies in industrialized construction. *Journal of Building Engineering*, *39*, 102265. https://doi.org/10.1016/j.jobe.2021.102265

Ramalingham, S., & Mossman, A. (2022). "Knowing with AI in construction - an empirical insight" *The 9th International Conference on Construction Engineering and Project Management* . https://www.researchgate.net/publication/362481839

Rathod, K., & Sonawane, A. (2022). Application of artificial intelligence in project planning to solve late and over-budgeted construction projects. *2022 International Conference on Sustainable Computing and Data Communication Systems (ICSCDS)*, 424–431. https://doi.org/10.1109/ICSCDS53736.2022.9761027

Regona, M., Yigitcanlar, T., Xia, B., & Li, R. Y. M. (2022). Opportunities and adoption challenges of AI in the construction industry: a prisma review. *Journal of Open*

Innovation: Technology, Market, and Complexity, *8*(1), 45. https://doi.org/10.3390/joitmc8010045

Skorupka, D. (2008). Identification and initial risk assessment of construction projects in Poland. *Journal of Management in Engineering*, *24*(3), 120–127. https://doi.org/10.1061/(ASCE)0742-597X(2008)24:3(120)

Son, H., Kim, C., Kim, H., Han, S. H., & Kim, M. K. (2010). Trend analysis of research and development on automation and robotics technology in the construction industry. *KSCE Journal of Civil Engineering*, *14*(2), 131–139. https://doi.org/10.1007/s12205-010-0131-7

Souhail, E., Elodie, H., M. Ammar, A., & Chance, E. (2020). *Will Artificial Intelligence (AI) Take over the Construction World? - A Multidisciplinary Exploration*. 59–66. https://doi.org/10.3311/ccc2020-010

Srivastava, A., Jawaid, S., Singh, R., Gehlot, A., Akram, S. V., Priyadarshi, N., & Khan, B. (2022). Imperative role of technology intervention and implementation for automation in the construction industry. *Advances in Civil Engineering*, *2022*, 1–19. https://doi.org/10.1155/2022/6716987

Srivastava, A., Rastogi, A., Rao, A., Shoeb, A. A. M., Abid, A., Fisch, A., Brown, A. R., Santoro, A., Gupta, A., & Garriga-Alonso, A. (2022). Beyond the imitation game: quantifying and extrapolating the capabilities of language models. *ArXiv Preprint ArXiv:2206.04615*.

Williams, N., P. Ferdinand, N., & Croft, R. (2014). Project management maturity in the age of big data. *International Journal of Managing Projects in Business*, *7*(2), 311–317. https://doi.org/10.1108/IJMPB-01-2014-0001

https://doi.org/10.52202/070182-0005

Zaman, U., Nawaz, S., & Nadeem, R. D. (2020). Navigating innovation success through projects. role of CEO transformational leadership, project management best practices, and project management technology quotient. *Journal of Open Innovation: Technology, Market, and Complexity*, *6*(4), 168. https://doi.org/10.3390/joitmc6040168

Biography

Gurneet Singh Sidhu is a master's student in Project Management at Northeastern University. He holds a Bachelor of Technology in Civil Engineering from Manipal University. His research delves into the influence of emerging technologies on project management, particularly within the construction sector. Focusing on the potential benefits and challenges of AI integration, Gurneet aims to optimize project execution, resource allocation, and decision-making processes in the construction industry. Additionally, he explores the merits of adopting ambidextrous leadership in the contemporary technological era.

Adel A. Zadeh is an Associate Teaching Professor at Northeastern University in Toronto in the Project Management program. With a PhD in construction materials and sustainability from the University of Cambridge and a LEED AP certification, Adel is well-equipped to tackle the pressing issues facing the construction industry today. His current research delves into the global sand crisis and the implementation of lean principles in the construction industry.

https://doi.org/10.52202/070182-0005

AUTHOR INDEX

Northeastern University Global Symposium on Leadership and Project Management
Boston, MA
USA

CURRAN ASSOCIATES INC.
proceedings.com

ISBN 978-1-7138-7680-9